RENEWALS 458-4574

DATE DUE

**WITHDRAWN
UTSA LIBRARIES**

GENOMICS in REGULATORY ECOTOXICOLOGY

Applications and Challenges

Other Titles from the Society of Environmental Toxicology and Chemistry (SETAC)

Population-Level Ecological Risk Assessment
Barnthouse, Munns, Sorensen, editors
2007

Effects of Water Chemistry on Bioavailability and Toxicity of Waterborne Cadmium, Copper, Nickel, Lead, and Zinc on Freshwater Organisms
Meyer, Clearwater, Doser, Rogaczewski, Hansen
2007

Ecosystem Responses to Mercury Contamination: Indicators of Change
Harris, Krabbenhoft, Mason, Murray, Reash, Saltman, editors
2007

Genomic Approaches for Cross-Species Extrapolation in Toxicology
Benson and Di Giulio, editors
2007

New Improvements in the Aquatic Ecological Risk Assessment of Fungicidal Pesticides and Biocides
Van den Brink, Maltby, Wendt-Rasch, Heimbach, Peeters, editors
2007

Freshwater Bivalve Ecotoxicology
Farris and Van Hassel, editors
2006

*Estrogens and Xenoestrogens in the Aquatic Environment:
An Integrated Approach for Field Monitoring and Effect Assessment*
Vethaak, Schrap, de Voogt, editors
2006

Assessing the Hazard of Metals and Inorganic Metal Substances in Aquatic and Terrestrial Systems
Adams and Chapman, editors
2006

Perchlorate Ecotoxicology
Kendall and Smith, editors
2006

Natural Attenuation of Trace Element Availability in Soils
Hamon, McLaughlin, Stevens, editors
2006

Mercury Cycling in a Wetland-Dominated Ecosystem: A Multidisciplinary Study
O'Driscoll, Rencz, Lean
2005

For information about SETAC publications, including SETAC's international journals, *Environmental Toxicology* and *Chemistry and Integrated Environmental Assessment and Management*, contact the SETAC Administrative Office nearest you:

SETAC Office
1010 North 12th Avenue
Pensacola, FL 32501-3367 USA
T 850 469 1500 F 850 469 9778
E setac@setac.org

SETAC Office
Avenue de la Toison d'Or 67
B-1060 Brussels, Belgium
T 32 2 772 72 81 F 32 2 770 53 86
E setac@setaceu.org

www.setac.org
Environmental Quality Through Science®

GENOMICS in REGULATORY ECOTOXICOLOGY

Applications and Challenges

Edited by
Gerald T. Ankley
Ann L. Miracle
Edward J. Perkins
George P. Daston

Coordinating Editor of SETAC Books
Joseph W. Gorsuch
Gorsuch Environmental Management Services, Inc.
Webster, New York, USA

CRC Press is an imprint of the
Taylor & Francis Group, an **informa** business

Published in collaboration with the Society of Environmental Toxicology and Chemistry (SETAC)
1010 North 12th Avenue, Pensacola, Florida 32501
Telephone: (850) 469-1500 ; Fax: (850) 469-9778; Email: setac@setac.org
Web site: www.setac.org
ISBN: 978-1-880611-38-8 (SETAC Press)

© 2008 by the Society of Environmental Toxicology and Chemistry (SETAC)
CRC Press is an imprint of Taylor & Francis Group, an Informa business

No claim to original U.S. Government works
Printed in the United States of America on acid-free paper
10 9 8 7 6 5 4 3 2 1

International Standard Book Number-13: 978-1-4200-6682-1 (Hardcover)

This book contains information obtained from authentic and highly regarded sources. Reprinted material is quoted with permission, and sources are indicated. A wide variety of references are listed. Reasonable efforts have been made to publish reliable data and information, but the author and the publisher cannot assume responsibility for the validity of all materials or for the consequences of their use. Information contained herein does not necessarily reflect the policy or views of the Society of Environmental Toxicology and Chemistry (SETAC). Mention of commercial or noncommercial products and services does not imply endorsement or affiliation by the author or SETAC.

The content of this publication does not necessarily reflect the position or policy of the U.S. government or sponsoring organizations and an official endorsement should not be inferred.

Except as permitted under U.S. Copyright Law, no part of this book may be reprinted, reproduced, transmitted, or utilized in any form by any electronic, mechanical, or other means, now known or hereafter invented, including photocopying, microfilming, and recording, or in any information storage or retrieval system, without written permission from the publishers.

For permission to photocopy or use material electronically from this work, please access www.copyright.com (http://www.copyright.com/) or contact the Copyright Clearance Center, Inc. (CCC) 222 Rosewood Drive, Danvers, MA 01923, 978-750-8400. CCC is a not-for-profit organization that provides licenses and registration for a variety of users. For organizations that have been granted a photocopy license by the CCC, a separate system of payment has been arranged.

Trademark Notice: Product or corporate names may be trademarks or registered trademarks, and are used only for identification and explanation without intent to infringe.

Library of Congress Cataloging-in-Publication Data

Genomics in Regulatory Ecotoxicology: Applications and Challenges / editor(s), Gerald Ankley, Ann Miracle.
  p. ; cm.
Includes bibliographical references and index.
ISBN 978-1-4200-6682-1 (hardback : alk. paper)
1. Ecological risk assessment--Methodology. 2. Genetic toxicology. 3. Environmental toxicology. 4. Genomics. I. Ankley, Gerald T. (Gerald Thomas), 1959- II. Miracle, Ann. III. SETAC (Society)
  [DNLM: 1. Toxicogenetics--methods. 2. Environmental Pollutants--toxicity. 3. Microarray Analysis--methods. 4. Risk Assessment. QV 38 T7547 2007]

QH541.15.R57T69 2007
577.27--dc22   2007033028

Visit the Taylor & Francis Web site at
http://www.taylorandfrancis.com

and the CRC Press Web site at
http://www.crcpress.com

and the SETAC Web site at
www.setac.org

Library
University of Texas
at San Antonio

SETAC Publications

Books published by the Society of Environmental Toxicology and Chemistry (SETAC) provide in-depth reviews and critical appraisals on scientific subjects relevant to understanding the impacts of chemicals and technology on the environment. The books explore topics reviewed and recommended by the Publications Advisory Council and approved by the SETAC North America, Latin America, or Asia/Pacific Board of Directors; the SETAC Europe Council; or the SETAC World Council for their importance, timeliness, and contribution to multidisciplinary approaches to solving environmental problems. The diversity and breadth of subjects covered in the series reflect the wide range of disciplines encompassed by environmental toxicology, environmental chemistry, hazard and risk assessment, and life-cycle assessment. SETAC books attempt to present the reader with authoritative coverage of the literature, as well as paradigms, methodologies, and controversies; research needs; and new developments specific to the featured topics. The books are generally peer reviewed for SETAC by acknowledged experts.

SETAC publications, which include Technical Issue Papers (TIPs), workshop summaries, newsletter (*SETAC Globe*), and journals (*Environmental Toxicology and Chemistry* and *Integrated Environmental Assessment and Management*), are useful to environmental scientists in research, research management, chemical manufacturing and regulation, risk assessment, and education, as well as to students considering or preparing for careers in these areas. The publications provide information for keeping abreast of recent developments in familiar subject areas and for rapid introduction to principles and approaches in new subject areas.

SETAC recognizes and thanks the past coordinating editors of SETAC books:

    A.S. Green, International Zinc Association
        Durham, North Carolina, USA

    C.G. Ingersoll, Columbia Environmental Research Center
        US Geological Survey, Columbia, Missouri, USA

    T.W. La Point, Institute of Applied Sciences
        University of North Texas, Denton, Texas, USA

    B.T. Walton, US Environmental Protection Agency
        Research Triangle Park, North Carolina, USA

    C.H. Ward, Department of Environmental Sciences and Engineering
        Rice University, Houston, Texas, USA

Contents

SETAC Publications v
List of Figures ix
List of Tables xi
Foreword xiii
Preface xvii
Acknowledgments xix
The Editors xxi
Contributors xxv

Chapter 1 Toxicogenomics in Ecological Risk Assessments: Regulatory Context, Technical Background, and Workshop Overview 1

  Gerald T Ankley, Ann L Miracle, and Edward J Perkins

Chapter 2 Application of Genomics to Screening 13

  *Sean W Kennedy, Susan Euling, Duane B Huggett,
  L Wim De Coen, Jason R Snape, Timothy R Zacharewski, and
  Jun Kanno*

Chapter 3 Application of Genomics to Tiered Testing 33

  *Charles R Tyler, Amy L Filby, Taisen Iguchi, Vincent J Kramer,
  DG Joakim Larsson, Graham van Aggelen, Kees van Leeuwen,
  Mark R Viant, and Donald E Tillitt*

Chapter 4 Application of Genomics to Regulatory Ecological Risk Assessments for Pesticides 63

  *Sigmund J Degitz, Robert A Hoke, Steven Bradbury,
  Richard Brennan, Lee Ferguson, Rebecca Klaper, Laszlo
  Orban, David Spurgeon, and Susan Tilton*

Chapter 5 Application of Genomics to Assessment of the Ecological Risk of Complex Mixtures 87

  *Edward J Perkins, Nancy Denslow, J Kevin Chipman,
  Patrick D Guiney, James R. Oris, Helen Poynton,
  Pierre Yves Robidoux, Richard Scroggins, and Glen Van Der Kraak*

Chapter 6 Application of Genomic Technologies to Ecological Risk
Assessment at Remediation and Restoration Sites............................ 123

*Ann L Miracle, Clive W Evans, Elizabeth A Ferguson,
Bruce Greenberg, Peter Kille, Anton R Schaeffner,
Mark Sprenger, Ronny van Aerle, and Donald J Versteeg*

Chapter 7 Toxicogenomics in Ecological Risk Assessments:
A Prospectus 151

*George P Daston, Ann L Miracle, Edward J Perkins, and
Gerald T Ankley*

Glossary 157

Index 161

List of Figures

FIGURE 3.1 Framework illustrating how combining use and exposure information and effects information obtained from (quantitative) structure–activity relationships ([Q]SARs), read-across methods, thresholds of toxicological concern (TTCs), and in vitro tests prior to in vivo testing can provide a more rapid, efficient, and cost-effective way to perform risk assessment of chemicals 37

FIGURE 3.2 Generic tiered-testing and risk assessment framework for prospective ecological risk assessments indicating where toxicogenomic data could be incorporated relative to other models. Rapid prioritization arrows indicate where genomics could be used to inform on hazard assessment and better direct higher tiers of testing 45

FIGURE 3.3 Conceptual timeline and relative resource requirements for the development and integration of toxicogenomic data into regulatory programs 48

FIGURE 5.1 Integration of genomics with monitoring in a tiered assessment. The ecological risk assessment framework is shown as a tiered process 90

FIGURE 5.2 The linkage between genomic responses and adverse outcomes of concern to regulators. Note that this also is the area of greatest uncertainty—extrapolating results down to low doses 103

FIGURE 5.3 Schematic of systems toxicology 113

FIGURE 6.1 Ecological risk assessment (ERA) framework 126

FIGURE 6.2 Areas within the characterization of ecological risk assessment in which genomic technologies can provide clarity within the remediation framework 132

List of Tables

TABLE 1.1 Plenary session for Pellston workshop entitled "Molecular Biology and Risk Assessment: Evaluation of the Potential Roles of Genomics in Regulatory Ecotoxicology" 7

TABLE 2.1 Priority research questions and recommendations for the development of genomics-based ecotoxicological screening assays 22

TABLE 2.2 Information required to assist in validation of assays for use in regulatory decision making.. 23

TABLE 4.1 Typical species utilized for toxicity testing in support of pesticide risk assessment.. 69

TABLE 4.2 Selected biomarkers for exposure to various types of chemicals 71

TABLE 4.3 Typical toxicity testing endpoints for aquatic and terrestrial species used in pesticide risk assessments... 73

TABLE 5.1 Examples of regulatory and status and trend environmental monitoring programs and tools 91

TABLE 6.1 Selected examples of the application of genomic technologies to remediation 135

TABLE 6.2 Specific genomic tools by molecular compartment that identify relevant application and how it has been used in environmental assessments 141

TABLE 6.3 Considerations for effective use of genomics technologies in the retrospective risk assessment process 144

Foreword

The workshop from which this book resulted, Molecular Biology and Risk Assessment: Evaluation of the Potential Roles of Genomics in Regulatory Ecotoxicology, held in Pensacola, Florida, 18-22 September 2005, was part of the successful "Pellston Workshop Series." Since 1977, Pellston Workshops have brought scientists together to evaluate current and prospective environmental issues. Each workshop has focused on a relevant environmental topic, and the proceedings of each have been published as peer-reviewed or informal reports. These documents have been widely distributed and are valued by environmental scientists, engineers, regulators, and managers for their technical basis and their comprehensive, state-of-the-science reviews. The other workshops in the Pellston series are as follows:

- Estimating the Hazard of Chemical Substances to Aquatic Life. Pellston, Michigan, 13–17 Jun 1977. Published by the American Society for Testing and Materials, STP 657, 1978.
- Analyzing the Hazard Evaluation Process. Waterville Valley, New Hampshire, 14–18 Aug 1978. Published by The American Fisheries Society, 1979.
- Biotransformation and Fate of Chemicals in the Aquatic Environment. Pellston, Michigan, 14–18 Aug 1979. Published by The American Society of Microbiology, 1980.
- Modeling the Fate of Chemicals in the Aquatic Environment. Pellston, Michigan, 16–21 Aug 1981. Published by Ann Arbor Science, 1982.
- Environmental Hazard Assessment of Effluents. Cody, Wyoming, 23–27 Aug 1982. Published as a SETAC Special Publication by Pergamon Press, 1985.
- Fate and Effects of Sediment-Bound in Aquatic Systems. Florissant, Colorado, 11–18 Aug 1984. Published as a SETAC Special Publication by Pergamon Press, 1987.
- Research Priorities in Environmental Risk Assessment. Breckenridge, Colorado, 16–21 Aug 1987. Published by SETAC, 1987.
- Biomarkers: Biochemical, Physiological, and Histological Markers of Anthropogenic Stress. Keystone, Colorado, 23–28 Jul 1989. Published as a SETAC Special Publication by Lewis Publishers, 1992.
- Population Ecology and Wildlife Toxicology of Agricultural Pesticide Use: A Modeling Initiative for Avian Species. Kiawah Island, South Carolina, 22–27 Jul 1990. Published as a SETAC Special Publication by Lewis Publishers, 1994.
- A Technical Framework for [Product] Life-Cycle Assessments. Smuggler's Notch, Vermont, 18–23 August 1990. Published by SETAC, Jan 1991; 2nd printing Sep 1991; 3rd printing Mar 1994.

- Aquatic Microcosms for Ecological Assessment of Pesticides. Wintergreen, Virginia, 7–11 Oct 1991. Published by SETAC, 1992.
- A Conceptual Framework for Life-Cycle Assessment Impact Assessment. Sandestin, Florida, 1–6 Feb 1992. Published by SETAC, 1993.
- A Mechanistic Understanding of Bioavailability: Physical–Chemical Interactions. Pellston, Michigan, 17–22 Aug 1992. Published as a SETAC Special Publication by Lewis Publishers, 1994.
- Life-Cycle Assessment Data Quality Workshop. Wintergreen, Virginia, 4–9 Oct 1992. Published by SETAC, 1994.
- Avian Radio Telemetry in Support of Pesticide Field Studies. Pacific Grove, California, 5–8 Jan 1993. Published by SETAC, 1998.
- Sustainability-Based Environmental Management. Pellston, Michigan, 25–31 Aug 1993. Co-sponsored by the Ecological Society of America. Published by SETAC, 1998.
- Ecotoxicological Risk Assessment for Chlorinated Organic Chemicals. Alliston, Ontario, Canada, 25–29 Jul 1994. Published by SETAC, 1998.
- Application of Life-Cycle Assessment to Public Policy. Wintergreen, Virginia, 14–19 Aug 1994. Published by SETAC, 1997.
- Ecological Risk Assessment Decision Support System. Pellston, Michigan, 23–28 Aug 1994. Published by SETAC, 1998.
- Avian Toxicity Testing. Pensacola, Florida, 4–7 Dec 1994. Co-sponsored by Organisation for Economic Co-operation and Development. Published by OECD, 1996.
- Chemical Ranking and Scoring (CRS): Guidelines for Developing and Implementing Tools for Relative Chemical Assessments. Sandestin, Florida, 12–16 Feb 1995. Published by SETAC, 1997.
- Ecological Risk Assessment of Contaminated Sediments. Pacific Grove, California, 23–28 Apr 1995. Published by SETAC, 1997.
- Ecotoxicology and Risk Assessment for Wetlands. Fairmont, Montana, 30 Jul–3 Aug 1995. Published by SETAC, 1999.
- Uncertainty in Ecological Risk Assessment. Pellston, Michigan, 23–28 Aug 1995. Published by SETAC, 1998.
- Whole-Effluent Toxicity Testing: An Evaluation of Methods and Prediction of Receiving System Impacts. Pellston, Michigan, 16–21 Sep 1995. Published by SETAC, 1996.
- Reproductive and Developmental Effects of Contaminants in Oviparous Vertebrates. Fairmont, Montana, 13–18 Jul 1997. Published by SETAC, 1999.
- Multiple Stressors in Ecological Risk Assessment. Pellston, Michigan, 13–18 Sep 1997. Published by SETAC, 1999.
- Re-evaluation of the State of the Science for Water Quality Criteria Development. Fairmont, Montana, 25–30 Jun 1998. Published by SETAC, 2003.
- Criteria for Persistence and Long-Range Transport of Chemicals in the Environment. Fairmont Hot Springs, British Columbia, Canada, 14–19 Jul 1998. Published by SETAC. 2000.

- Assessing Contaminated Soils: From Soil-Contaminant Interactions to Ecosystem Management. Pellston, Michigan, 23–27 Sep 1998. Published by SETAC, 2003.
- Endocrine Disruption in Invertebrates: Endocrinology, Testing, and Assessment (EDIETA). Amsterdam, The Netherlands, 12–15 Dec 1998. Published by SETAC, 1999.
- Assessing the Effects of Complex Stressors in Ecosystems. Pellston, Michigan, 11–16 Sep 1999. Published by SETAC, 2001.
- Environmental–Human Health Interconnections. Snowbird, Utah, 10–15 Jun 2000. Published by SETAC, 2002.
- Ecological Assessment of Aquatic Resources: Application, Implementation, and Communication. Pellston, Michigan, 16–21 Sep 2000. Published by SETAC, 2004.
- The Global Decline of Amphibian Populations: An Integrated Analysis of Multiple Stressors Effects. Wingspread, Racine, Wisconsin, 18–23 Aug 2001. Published by SETAC, 2003.
- Methods of Uncertainty Analysis for Pesticide Risks. Pensacola, Florida, 24 Feb–1 Mar 2002.
- The Role of Dietary Exposure in the Evaluation of Risk of Metals to Aquatic Organisms. Fairmont Hot Springs, British Columbia, Canada, 27 Jul–1 Aug 2002. Published by SETAC, 2005.
- Use of Sediment Quality Guidelines (SQGs) and Related Tools for the Assessment of Contaminated Sediments. Fairmont Hot Springs, Montana, 17–22 Aug 2002. Published by SETAC, 2005.
- Science for Assessment of the Impacts of Human Pharmaceuticals on Aquatic Ecosystem. Snowbird, Utah, 3–8 Jun 2003. Published by SETAC, 2005.
- Valuation of Ecological Resources: Integration of Ecological Risk Assessment and Socio-Economics to Support Environmental Decisions. Pensacola, Florida, 4–9 Oct 2003. Published by SETAC and CRC Press, 2007.
- Emerging Molecular and Computational Approaches for Cross-Species Extrapolations. Portland, Oregon, 18–22 Jul 2004. Published by SETAC and CRC Press, 2006.
- Veterinary Pharmaceuticals. Pensacola, Florida, 12–16 Feb 2006. To be published by SETAC and CRC Press, 2008.

Preface

The workshop that served as the basis of this book was held 18–22 September 2005, in Pellston, Michigan, organized by the Society of Environmental Toxicology and Chemistry (SETAC). One goal of the workshop was to develop a conceptual framework concerning how genomic data could have effective short- and long-term impacts on different types of ecological risk assessments. An additional goal was to make recommendations to guide genomic research such that future developments would have a greater likelihood of contributing to regulatory decision making. Pellston Workshops have a long and successful history of aligning environmental science and regulation, so it was felt that this is would be an ideal venue for addressing the topic. Participants at the workshop included both technical experts in the area of genomics and risk assessment experts who potentially could use genomic data in regulatory activities.

Recent years have witnessed a proliferation of different genomic approaches that, through a combination of advanced biological, instrumental, and informatic techniques, can yield a previously unparalleled amount of data concerning the molecular and biochemical status of organisms. Fueled partially by large, well-publicized efforts such as the Human Genome Project, genomic research has become a rapidly growing topical area in multiple biological disciplines, including toxicology.

There has been considerable discussion of the potential utility of genomics in addressing data gaps and uncertainties in the arena of regulatory toxicology. Regulatory needs for improved safety data for both human and ecological risk assessments are increasing rapidly. New testing efforts such as the REACH (registration, evaluation, authorization, and restriction of chemicals) program within the European Union and the high-production volume (HPV) challenge program in the United States promise to increase significantly the number of chemicals for which toxicity data may be needed. Emerging pollutants of concern, such as human and veterinary pharmaceuticals, also are anticipated to increase regulatory testing requirements for ecological effects.

Against this backdrop of increased testing, the registration of new chemicals and the re-registration of existing materials (e.g., pesticides) continue. In addition, there is a steady increase in the frequency of diagnostic assessments that rely on toxicity tests with complex environmental samples, such as ambient waters, effluents, and sediments, as a basis for compliance monitoring and/or evaluation of treatment and remediation efforts. The complexity of testing also is increasing. For example, regulatory programs throughout the world are starting to incorporate tests and endpoints that capture the effects of chemicals with the potential to disrupt specific endocrine pathways in animals. Overall, therefore, ecological testing and screening programs need to be more thorough, less costly, and able to be implemented more rapidly.

At present, virtually all methods used as a basis for regulatory decision making in ecotoxicology rely on whole-animal exposures that focus on effects on survival, growth, and reproduction. These types of tests are resource and time intensive—especially as one moves from the realm of short-term lethality assays to partial and full life-cycle tests. Given the anticipated amount of new testing, the time required to accomplish this will be measured in decades unless approaches can be developed to streamline the testing process.

Toxicogenomic tools offer the potential to effectively address a number of the data needs and uncertainties currently confronting risk assessors and regulators. However, exactly how toxicogenomics might be incorporated into regulatory programs is uncertain. The goal of the SETAC Pellston Workshop described in this book was to define how this challenge could be met.

Acknowledgments

This publication is a result of a SETAC workshop held in Pellston, Michigan, 18–22 September 2005. Support for the workshop was provided by AstraZeneca (UK); Dow AgroSciences, LLC; EI DuPont deNemours Co.; Environment Canada; GSF-National Research Center for Environment and Health (Germany); Miami University of Ohio; UK Natural Environmental Research Council; Pfizer, Inc.; SC Johnson and Son, Inc.; University of Antwerp (Belgium); UK Environment Agency; US Army Corps of Engineers Research and Development Center; and US Environmental Protection Agency. The commitment of the sponsors to advancing environmental sciences is appreciated.

Sincere thanks are expressed to all technical contributors to this document, many of whom spent countless hours preparing and reviewing the various chapters. Special thanks are offered to the SETAC personnel who organized and conducted the workshop (Greg Schiefer, Taylor Mitchell, and Jason Andersen) and those who played a role in the editorial preparation of this book (Mimi Meredith and the volunteer coordinating editor of SETAC books, Joe Gorsuch). Dr Daniel Villeneuve reviewed an earlier version of the book and provided comments that substantially improved the final manuscript.

The Editors

Gerald T Ankley is a research toxicologist with the US Environmental Protection Agency (USEPA) Mid-Continent Ecology Division in Duluth, Minnesota. Dr Ankley received his BS from the Department of Fisheries and Wildlife at Michigan State University, and his MS and PhD degrees from the School of Forest Resources at the University of Georgia. He has worked at the Duluth USEPA lab for about 20 years in a variety of areas, including the development of test methods for effluents and sediments, determination of the direct and indirect impacts of solar ultraviolet radiation on aquatic organisms, assessment of the effects of endocrine-disrupting chemicals on wildlife, and, most recently, application of genomic and computational toxicology tools to ecological risk assessments. He has authored more than 225 research papers on these and related topics and recently was recognized as one of the most widely cited scientists in the world in ecology and environmental sciences. Dr Ankley is involved in a number of national and international organizations focused on the application of state-of-the-art science to regulatory ecotoxicology; he has been an active member of SETAC for about 25 years.

Ann L Miracle began her training in molecular ecology with the US Environmental Protection Agency, where she developed microarrays for the fathead minnow and led multilaboratory projects that involved incorporating genomics, proteomics, and metabolite profiling to understand the toxicity of various endocrine-disrupting compounds with different modes of action. Dr Miracle is currently employed with the Pacific Northwest National Laboratory as a senior scientist; she leads the Predicting Ecosystem Change and Damage focus area of the Environmental Biomarkers Initiative that addresses phylogenetically diverse periphyton and microbial community responses to environmental contaminants. Her current research interests involve the incorporation of environmental biomarkers into relevant risk assessment guidelines and investigation of fate and transport of nanomaterials in ecological models.

Edward J Perkins is a research biologist and team leader of the Environmental Genomics and Genetics Team in US Army Engineer Research and Development Center (ERDC), Environmental Laboratory. Dr Perkins received his PhD studying biodegradation of 2, 4-dichlorophenoxyacetic acid (24D) from Washington State University in genetics and cell biology in 1987. Dr Perkins continued studying the genetics of 24D biodegradation at the University of Washington, followed by research into molecular measures of soil quality at the USDA Research Station in Pullman, Washington. He joined the ERDC Environmental Laboratory in 1996 where he established a genetics research lab. His laboratory utilizes genomics to examine the effect of military activities on ecologically important organisms in an effort to minimize adverse environmental impacts on Department of Defense lands.

George P Daston is a research fellow at Procter & Gamble where he has worked for more than 21 years. He has spent his entire career researching the effects of exogenous chemicals on living systems, especially the developing embryo, fetus, and child. He has published more than 100 peer-reviewed articles, reviews, and book chapters and has edited 3 books. His most recent research includes genomic approaches to endocrine disrupter screening and improvements in risk assessment methodology. Dr Daston's activities in professional societies include serving as chair of the Developmental and Reproductive Toxicology Technical Committee of ILSI-Health Effects Sciences Institute (1996–2004); president of the Society of Toxicology's Reproductive and Developmental Toxicology Specialty Section (1994–1995); president of the Teratology Society (1999–2000); member of the National Academy of Sciences Board on Environmental Studies and Toxicology (1995–1998); councilor of the Society of Toxicology (2001–2003); member of the USEPA Board of Scientific Counselors; member of the National Toxicology Program Board of Scientific Counselors; and member of the NIH National Children's Study Advisory Committee.

Dr Daston is editor-in-chief of *Birth Defects Research: Developmental and Reproductive Toxicology*, on the editorial board of *Human and Ecological Risk Assessment*, and an ad hoc reviewer for *Journal of Nutrition, FASEB Journal*, and other journals. He served for 6 years as associate editor of *Toxicologcal Sciences*. Dr Daston is an adjunct professor in the Department of Pediatrics and Developmental Biology Program at the University of Cincinnati and Children's Hospital Research Foundation. He was a visiting scientist at the Salk Institute, Molecular Neurobiology Laboratory (1993–1994). Dr Daston was elected a fellow of AAAS in 1999 and of the Academy of Toxicological Sciences in 2005.

Contributors

Gerald T Ankley
US Environmental Protection Agency
Duluth, Minnesota

Steven Bradbury
US Environmental Protection Agency
Washington, DC

Richard Brennan
Iconix Pharmaceuticals, Inc.
Mountain View, California

J Kevin Chipman
The University of Birmingham
Birmingham, United Kingdom

George P Daston
The Procter & Gamble Company
Cincinnati, Ohio

L Wim De Coen
University of Antwerp
Antwerp, Belgium

Sigmund J Degitz
US Environmental Protection Agency
Duluth, Minnesota

Nancy Denslow
University of Florida
Gainesville, Florida

Susan Euling
US Environmental Protection Agency
Washington, DC

Clive W Evans
University of Auckland
Auckland, New Zealand

Elizabeth A Ferguson
US Army Corps of Engineers
Vicksburg, Mississippi

Lee Ferguson
University of South Carolina
Columbia, South Carolina

Amy L Filby
The University of Exeter
Exeter, United Kingdom

Bruce Greenberg
University of Waterloo
Waterloo, Ontario, Canada

Patrick D Guiney
SC Johnson & Son, Inc.
Racine, Wisconsin

Robert A Hoke
EI Dupont de Nemours Company
Newark, Delaware

Duane B Huggett
Pfizer, Inc.
Groton, Connecticut

Taisen Iguchi
National Institutes of Natural Science
Okazaki, Aichi-ken, Japan

Jun Kanno
National Institutes of Health Sciences
Tokyo, Japan

Sean W Kennedy
Environment Canada
Ottawa, Ontario, Canada

Peter Kille
Cardiff University
Cardiff, United Kingdom

Rebecca Klaper
University of Wisconsin-Milwaukee
Milwaukee, Wisconsin

Vincent J Kramer
Dow AgroSciences LLC
Indianapolis, Indiana

DG Joakim Larsson
Goteborg University
Goteborg, Sweden

Ann L Miracle
Pacific Northwest National Laboratory
Richland, Washington

Laszlo Orban
National University of Singapore
Singapore, Singapore

James R Oris
Miami University
Oxford, Ohio

Edward J Perkins
US Army Engineer
Research and Development Center
Vicksburg, Mississippi

Helen Poynton
University of California-Berkeley
Berkeley, California

Pierre Yves Robidoux
National Research Council Canada
Montreal, Quebec, Canada

Anton R Schaeffner
GSF Institute of Biochemical Plant
  Pathology
Munich, Germany

Richard Scroggins
Environment Canada
Gloucester, Ontario, Canada

Jason R Snape
AstraZeneca
Brixham, United Kingdom

Mark Sprenger
US Environmental Protection Agency
Edison, New Jersey

David Spurgeon
Monks Wood
Huntingdon, United Kingdom

Donald E Tillitt
US Geological Survey
Columbia, Missouri

Susan Tilton
University of Washington
Seattle, Washington

Charles R Tyler
The University of Exeter
Exeter, United Kingdom

Ronny van Aerle
The University of Exeter
Exeter, United Kingdom

Graham van Aggelen
Environment Canada
North Vancouver, British Columbia,
  Canada

Glen Van Der Kraak
University of Guelph
Guelph, Ontario, Canada

Kees van Leeuwen
European Commission
Ispra, Italy

Donald Versteeg
The Procter & Gamble Company
Cincinnati, Ohio

Mark R Viant
University of Birmingham
Birmingham, United Kingdom

Timothy R Zacharewski
Michigan State University
East Lansing, Michigan

1 Toxicogenomics in Ecological Risk Assessments: Regulatory Context, Technical Background, and Workshop Overview

Gerald T Ankley, Ann L Miracle, and Edward J Perkins

CONTENTS

1.1 Regulatory Context 2
1.2 Technical Background 3
1.3 Workshop Overview 5
    1.3.1 Scope of the Workshop: Global versus Focused Molecular Analyses 6
    1.3.2 Workshop Structure 7
        1.3.2.1 Workgroup 1: Application of Genomics to Screening-Level Risk Assessments 8
        1.3.2.2 Workgroup 2: Application of Genomics to Tiered Testing 8
        1.3.2.3 Workgroup 3: Application of Genomics to Comprehensive Assessments 8
        1.3.3.4 Workgroup 4: Application of Genomics to Environmental Monitoring 9
        1.3.3.5 Workgroup 5: Application of Genomics to Environmental Remediation or Resource Recovery 9
1.4 Conclusion 9
References 9

1.1 REGULATORY CONTEXT

Recent years have witnessed a proliferation of different genomic approaches that, through a combination of advanced biological, instrumental, and informatic techniques, can yield a previously unparalleled amount of data concerning the molecular and biochemical status of organisms. Fueled partially by large, well-publicized efforts such as the Human Genome Project, genomic research has become a rapidly growing topical area in multiple biological disciplines, including toxicology (Nuwaysir et al. 1999).

There has been considerable discussion of the potential utility of genomics in addressing data gaps and uncertainties in the arena of regulatory toxicology (MacGregor 2003; Waters and Fostel 2004; Boverhof and Zacharewski 2005; Luhe et al. 2005; Ankley et al. 2006). Regulatory needs for improved safety data for both human and ecological risk assessments are increasing rapidly. New testing efforts such as the REACH (registration, evaluation, authorization, and restriction of chemicals) program within the European Union and the high production volume (HPV) challenge program in the United States promise to increase significantly the number of chemicals for which toxicity data may be needed (European Commission 2001; www.epa.gov/chemrtk/volchall.htm). Emerging pollutants of concern, such as human and veterinary pharmaceuticals, also are anticipated to increase regulatory testing requirements for ecological effects (Ankley et al. 2005).

Against this backdrop of increased testing, the registration of new chemicals and the re-registration of existing materials (e.g., pesticides) continue. In addition, there is a steady increase in the frequency of diagnostic assessments that rely on toxicity tests with complex environmental samples, such as ambient waters, effluents, and sediments, as a basis for compliance monitoring and/or evaluation of treatment and remediation efforts. The complexity of testing also is increasing. For example, regulatory programs throughout the world are starting to incorporate tests and endpoints that capture the effects of chemicals with the potential to disrupt specific endocrine pathways in animals (WHO 2002). Overall, therefore, ecological testing and screening programs need to be more thorough, less costly, and able to be implemented more rapidly.

At present, virtually all methods used as a basis for regulatory decision making in ecotoxicology rely on whole-animal exposures that focus on effects on survival, growth, and reproduction. These types of tests are resource and time intensive—especially as one moves from the realm of short-term lethality assays to partial and full life-cycle tests. Given the anticipated amount of new testing, the time required to accomplish this will be measured in decades unless approaches can be developed to streamline the testing process.

Toxicogenomic tools offer the potential to effectively address a number of the data needs and uncertainties currently confronting risk assessors and regulators. However, exactly how toxicogenomics might be incorporated into regulatory programs is uncertain. The goal of the Society of Environmental Toxicology and Chemistry (SETAC) Pellston workshop described in this book was to define how this challenge could be met.

1.2 TECHNICAL BACKGROUND

The purpose of the following section is to provide a general overview concerning the collection and nature of the types of genomic data addressed in this book. Additional technical detail on many of these topics is supplied in subsequent chapters. As in any comparatively new field, variations in terminology and its usage exist in the extant literature. The Glossary presents and defines key terms for the purposes of this book.

The genome is the DNA sequence of an organism, and there is an ever-increasing number of species for which the entire genome has been elucidated (e.g., humans, rats, mice, zebrafish, *Caenorhabditis elegans*). Partial genome data exist for many others, including several species directly relevant to ecotoxicology—for example, rainbow trout (Thorgaard et al. 2002), Japanese medaka (Naruse et al. 2004), fathead minnow (Ankley and Villeneuve 2006), *Xenopus* sp. (http://genome.jgi-psf.org/Xentr4/Xentr4.home.html), and *Daphnia pulex* (Colbourne et al. 2005). Among other applications, genomic information can be used to design microarrays or gene chips for some (or all) of the genes in an organism. These can be used to determine which genes are up- or down-regulated (as transcribed mRNA) in a cell, tissue, organ, or organism under specific physiological conditions or in response to an environmental perturbation, such as exposure to a toxic chemical. The global detection and analysis of gene expression in this fashion is termed "transcriptomics."

Virtually all responses to external stressors, including toxicants, involve changes in normal patterns of gene expression (Nuwaysir et al. 1999; Merrick and Bruno 2004). Some of these responses are a direct result of the chemical, such as gene expression that is directed when a steroid hormone (or analog) binds to a transcription factor (receptor), thereby forming a complex that modulates (enhances or depresses) transcription of specific genes. Other responses to toxic chemicals are compensatory, in that they reflect adaptive responses of the organism to molecular damage or cellular dysfunction. Importantly, different mechanisms of toxicity can generate specific patterns of gene expression reflective of mechanism or mode of action (Merrick and Bruno 2004).

The number of genes needed to reflect a mechanism or mode of action will vary in a pathway-specific manner. However, given the large number of genes that can be queried using microarrays (several thousand for some species), this is not likely to be a limiting factor in the application of transcriptomics to toxicity research (e.g., Amin et al. 2002; Hamadeh et al. 2002). Microarray research with mammalian models has been far more prevalent than in species used in regulatory ecotoxicology. However, examples of transcriptomic tools and research with ecologically relevant species are increasing (e.g., Larkin et al. 2003; Miracle et al. 2003; Rasooly et al. 2003; Williams et al. 2003; Kimura et al. 2004; Snape et al. 2004; Tilton et al. 2005; van der Ven et al. 2005; von Schalburg et al. 2005; Lettieri 2006).

Transcription of mRNA is only an intermediate step in conversion of genetic information into proteins, the biochemical basis of biological function. Not all mRNA sequences are transcribed, and many proteins are modified (e.g., by phosphorylation, post-translational cleavage, etc.) before becoming physiologically active (Pennington et al. 1997; Fields 2001; Handam and Righetti 2003). Consequently, proteomics, the global evaluation of proteins, provides additional critical insights into biological pathways. Alterations in protein profiles can be used, in conjunction with transcriptomics, to understand responses of an organism to toxicants (Waters and Fostel 2004). Like transcriptomics, there has been a rapid evolution of proteomic methods capable of providing broad characterizations of proteins expressed within cells, organs, or, in some instances, whole organisms, including species relevant to ecotoxicology (Shrader et al. 2003; Stentiford et al. 2005). Methods vary, but they typically include protein isolation and separation steps with techniques like two-dimensional gel electrophoresis or high-pressure liquid chromatography, followed by mass spectral (MS) analyses to identify peptide profiles or amino acid sequence as a basis for identification of specific proteins (Aebersold and Mann 2003; Fountoulakis 2004; Silva and Geromanos 2005).

Metabolomics describes the global characterization of low molecular weight metabolites involved in all the biological reactions required for growth, maintenance, and normal function (Nicholson et al. 2002; Schmidt 2004; Robertson 2005). The metabolome includes a variety of polar organic compounds (e.g., amino acids, small peptides, glucosides), comparatively nonpolar molecules such as lipids, and even inorganic chemicals. Metabolomics could be thought of as a sophisticated version of traditional tests used for disease states, where endogenous metabolite profiles can be used as a diagnostic tool (Nicholson et al. 2002). As such, metabolomics captures a more integrated assessment of the physiological state of an organism than transcriptomics do or proteomics (Robertson 2005). Most metabolomic research to date has focused on aspects of human health (humans, rats, mice); however, recent work in the area has successfully utilized animals, including aquatic species, relevant to ecological risk assessments (e.g., Viant 2003; Viant et al. 2003; Bundy et al. 2004). Different high-resolution MS and nuclear magnetic resonance (NMR) techniques provide the primary basis for generating metabolomic data (Dunn et al. 2005).

Transcriptomic, proteomic, and metabolomic data are generated at different biological levels of organization and thus provide different insights as to the biochemical and molecular status of an organism. However, all 3 approaches have excellent potential for defining toxicity mechanisms and their relationship to adverse outcomes, particularly if used in a complementary manner (Ideker et al. 2001; MacGregor 2003; Robertson 2005). These techniques also have many parallel challenges with regard to data collection, integration, and interpretation. For example, none of them would be considered routine relative to the biological endpoints typically measured and used for the environmental regulation of chemicals: survival, growth and development, and reproduction.

Advanced expertise, reagents, and sometimes costly equipment are required to collect genomic data. Furthermore, advanced expertise and capabilities also are needed for data analysis. Due to the amount of information generated, analysis of toxicogenomic data requires sophisticated bioinformatic (or chemometric) approaches that enable consideration of possible changes in thousands of data points per sample (Hess et al. 2001; Waters and Fostel 2004; Yu et al. 2004; Robertson 2005). For example, in humans, there are estimates of 30 000, 100 000 (or more), and between 2000 and 20 000 components of the transcriptome, proteome, and metabolome, respectively (Schmidt 2004). Ecotoxicologists historically have seldom dealt with data sets of this magnitude, so current infrastructure in terms of computing facilities and training requires expansion to enable meaningful analysis of genomic data.

A final challenge relative to use of toxicogenomic techniques for ecotoxicology research or regulation involves knowledge of what exactly is changing when a treatment causes alterations in gene, protein, or metabolite expression. The genomes of most species traditionally used for regulatory ecotoxicology have not been characterized to the point where more than a comparative handful of genes (and/or translated proteins) is well understood in terms of identity (or function). This is somewhat in contrast to the situation for toxicologists involved in human health research, where the genomes of many experimental models (as well as humans) have been extensively characterized. Even the zebrafish, which has a relatively well-characterized genome, lacks robust annotation for many gene products (http://www.sanger.ac.uk/Projects/D_rerio).

Hence, even though tools exist to detect changes in transcriptomic or proteomic data in a variety of nonmammalian species, including many used for regulatory ecotoxicology, the baseline information needed to interpret these data fully is often lacking. Metabolomic data have similar limitations, albeit from a slightly different perspective. Specifically, as opposed to genes and/or proteins, endogenous metabolites may be less likely to exhibit a high degree of species specificity (although actual profiles will vary). However, software and libraries available for identifying specific metabolic products from NMR or MS spectra are not yet extensive enough to exhaustively probe analytical data that may represent hundreds to thousands of unique molecules.

Ongoing research will, in the long term, serve to obviate limitations related to the global identification of gene products, proteins, and metabolites in test species relevant to ecological risk assessments; for the nearer term, there are approaches that would still allow toxicogenomic data to be used in certain applications. For example, identification of important genes, proteins, or metabolites associated with specific pathways of concern can be done in a focused fashion for key test species, thereby enabling development of molecular "profiles" indicative of toxic mode or mechanism of action. Alternatively, "fingerprinting" techniques also could be used to support aspects of regulatory decision making in ecotoxicology. These techniques rely on patterns of responses in a particular suite of analytes to provide insights as to toxicity mechanisms, rather than identification of the specific components that have changed. Fingerprints (also referred to as "signatures") obtained from unknown chemicals then can be compared to profiles generated from chemical exposures to toxicants with established mechanisms to serve as a basis for identifying potential toxicity pathways (e.g., Amin et al. 2002; Bushel et al. 2002; Naciff et al. 2002; Johnson et al. 2003; Merrick and Bruno 2004; Fielden et al. 2005).

1.3 WORKSHOP OVERVIEW

The workshop that served as the basis of this book was held 18–22 September 2005, in Pellston, Michigan. The primary goal of the workshop was to develop a conceptual framework concerning how genomic data could effectively impact different types of ecological risk assessments in both the short and long term. An additional goal was to make recommendations to guide genomic research such that future developments would have a greater likelihood of contributing to regulatory decision making. Pellston workshops have a long and successful history of aligning environmental science and regulation, so it was felt that this would be an ideal venue for addressing this topic. Participants at the workshop included both technical experts in the area of genomics and risk assessment experts who potentially could use genomic data in regulatory activities (see list of contributors, xxv–xxvii).

1.3.1 Scope of the Workshop: Global versus Focused Molecular Analyses

Although the primary motivation for holding the workshop involved the application of newer "global" genomic tools to regulatory ecotoxicology, the role of molecular and biochemical endpoints that measure single gene or protein expression also was considered. For example, we considered the use of techniques such as polymerase chain reaction (PCR) assays or enzyme-linked immunosorbent assays to measure, respectively, single mRNA products or proteins diagnostic of exposure and effects of chemicals with well-defined modes or mechanisms of action.

There were pragmatic reasons for including these types of endpoints as part of the workshop. Specifically, the approaches used by ecotoxicologists over the past 20 years in the development, validation, and implementation of molecular and biochemical indicators of exposure and effects (commonly termed "biomarkers"; SETAC 1992) serve as a basis for comparable research that will need to be conducted to support the use of genomic data for ecological risk assessment. Importantly, it is possible to view movement from indicators of single to multiple gene responses as a continuum, with many of the same needs and challenges relative to their use in regulatory applications. For example, for any molecular or biochemical endpoint to be used effectively in regulatory testing—whether it involves one or thousands of gene products, method standardization and validation are required to ensure that technically defensible and robust data are generated.

Similarly, for some regulatory applications, it is important to understand linkages of endpoints across biological levels of organization—that is, to provide phenotypic anchoring of changes at molecular or biochemical levels to the adverse outcomes reflected in the apical endpoints historically used for toxicity tests (SETAC 1992). Both method validation and research concerning linkages across biological levels of organization are more advanced for single gene or protein responses than for data generated using genomic techniques. Some testing programs that, in the past, have seldom considered endpoints other than apical responses are starting to consider these types of single "biomarker" responses in the decision-making process. A recent example of this involves programs in the United Kingdom that are evaluating induction of the protein vitellogenin in fish as a valid indicator of undesirable levels of contamination by estrogenic chemicals (Hutchinson et al. 2006).

Hence, acceptance and use of single gene or protein biomarkers by risk assessors represents an important first step in the process toward using the greater amounts of molecular and biochemical data generated by genomic tools. Also, it is probable that the more data-rich genomics techniques (e.g., microarrays) will serve as the basis for

TABLE 1.1
Plenary Session[a] for Pellston Workshop Entitled "Molecular Biology and Risk Assessment: Evaluation of the Potential Roles of Genomics in Regulatory Ecotoxicology"

8:00 AM	Steve Bradbury, "Ranking and prioritizing chemical risk assessments: possible roles for molecular data to focus questions, information needs, and resources"
8:30 AM	Kees Van Leeuwen, "Regulatory needs from the EU perspective"
9:00 AM	Elizabeth Ferguson, "Regulatory needs for decision making with complex environmental mixtures: sediments and effluents"
9:30 AM	Glen Van Der Kraak, "Use of biomarkers in assessing effluents: linkage to adverse outcomes and other lessons learned"
10:00 AM	Duane Huggett, "Perspectives on the application of genomics to regulation of ecological effects of pharmaceuticals"
10:30 AM	Break
10:45 AM	George Daston, "Examples of the application of genomics to human health risk assessments/regulation"
11:15 AM	Kevin Chipman, "Biomarkers from (eco)toxicogenomics and the challenge of distinguishing between compensatory versus toxic responses"
11:45 AM	Peter Kille, "Examples of the application of genomics to terrestrial risk assessments"
12:15 PM	Lunch
1:15 PM	Lee Ferguson, "Proteomics and examples of potential uses of in eco risk assessment"
1:45 PM	Mark Viant, "What can metabolomics offer environmental risk assessment?"

[a] 19 September 2005.

discovery of "new" biomarkers that, for some regulatory applications, would be most efficiently measured using single gene expression approaches (e.g., PCR). Therefore, it seems probable that the application of molecular and biochemical data to regulatory decision making will involve iterations and interactions between techniques that measure single versus multiple endpoints.

1.3.2 Workshop Structure

The workshop was initiated with a plenary session (Table 1.1) with 3 purposes: 1) provide succinct overviews as to the current state of the art in the areas of transcriptomics, proteomics, and metabolomics; 2) inform participants with examples of how genomic data, to date, have been used for research and/or decisions concerning ecological risk; and 3) provide attendees with details concerning different types of risk assessment scenarios (as reflected by the 5 workgroups described next). Following the plenary session, scientists at the meeting participated in one of the following 5 workgroups, each of which was charged with identifying how genomic data could be used to enhance different aspects of ecological risk assessments. There was, of course, some degree of overlap in the topics considered by the workgroups; however, having multiple teams of experts viewing the same challenge from slightly different perspectives was viewed as desirable.

1.3.2.1 Workgroup 1: Application of Genomics to Screening-Level Risk Assessments

This workgroup addressed the requirements of different types of ecological risk assessments that could be labeled as screening and/or prioritization. Basically, in these types of assessments, minimal data are required or available for regulatory decision making. An example of this from a US Environmental Protection Agency (USEPA) perspective is the level of information available for assessing potential risks of most chemicals in the Toxic Substances Control Act (TSCA) inventory of more than 70 000 chemicals. Inexpensive, rapid chemical screening-level tests based on genomics would be one approach to improving the quality of these types of assessments. An important focus of this workgroup was the use of transcript profiles or fingerprints from in vitro and/or short-term in vivo studies to identify possible toxic modes of action.

1.3.2.2 Workgroup 2: Application of Genomics to Tiered Testing

A number of regulatory programs rely upon tiered testing to guide data collection for final assessments of risk. This testing theoretically should optimize testing resources by eliminating a subset of chemicals from further testing relatively early in the assessment and prioritizing the remaining chemicals for subsequent work. For example, both national and international programs designed to detect potential human health and ecological effects of endocrine-disrupting chemicals have been and are being designed to utilize initial tests to dictate where subsequent, more elaborate (and costly) testing is needed. Genomic data clearly could be used as a basis for advancing from tier to tier, but how this might be done is uncertain. One focus of this workgroup concerned the use of genomics to provide mechanistic data that can be "phenotypically anchored" to adverse outcomes in whole-animal tests. This workgroup also addressed the types and uses of different genomic tools as one moves through increasingly resource-intensive steps of the tiered assessments.

1.3.2.3 Workgroup 3: Application of Genomics to Comprehensive Assessments

Certain regulatory scenarios presuppose a significant amount of data generation and collection to support a risk assessment decision. For example, pesticide registration typically costs millions of dollars (US) to generate the complete data sets considered necessary for ecological risk assessments. It is possible to envision the use of genomic data to help better focus this type of testing. For example, genomic data may help identify which and/or how many taxonomic groups may be required to ensure that environmental quality guidelines are indeed protective. Hence, focal areas of this group concerned use of genomics for the identification of test species and endpoints necessary for technically defensible risk assessments, as well as the use of genomics to help reduce uncertainties associated with interspecies extrapolation and dose- or concentration-response analysis (e.g., high- to low-dose extrapolation).

1.3.3.4 Workgroup 4: Application of Genomics to Environmental Monitoring

Several regulatory programs rely on data collected from biologically based environmental monitoring to make decisions. Examples include disposal of dredged material and permitting for effluents, where responses in animals collected from the field and/or exposed to field samples in the lab are used to ascertain the potential for chemical exposure and adverse effects. Use of genomics in these scenarios could have diagnostic potential in terms of addressing complex mixtures of chemical and nonchemical stressors; it also could provide sensitive endpoints predictive of potential adverse outcomes before they are manifested in populations. As such, this group considered the use of genomic tools in scenarios where complex mixtures of stressors occur, as well as the sensitivity of genomic endpoints relative to adverse apical responses.

1.3.3.5 Workgroup 5: Application of Genomics to Environmental Remediation or Resource Recovery

A final regulatory arena where genomic data could prove useful involves decisions concerning sites with existing impacts. Genomics could be useful for identifying site-specific stressors of most concern as a basis for determining remedial responses. Further, genomic data could serve as a useful method for evaluating recovery in conjunction with ongoing remediation. Focus areas in this workgroup were in some cases similar to those in Workgroup 4, but also included aspects of interspecies extrapolation (particularly, for example, when endangered or threatened species may be involved).

1.4 CONCLUSION

Interspersed with the approximately 2 days of workgroup meetings were periodic plenary sessions during which all the workshop participants assembled to discuss progress and address questions arising during discussions. Drafts of each chapter were completed at the meeting; over the course of the following several months and through multiple iterations of writing and review by those involved in the workshop, they were developed into the finished products contained in this book. A brief overview of the workshop was published in the open literature within about 10 months of completion of the meeting (Ankley et al. 2006).

REFERENCES

Aebersold R, Mann M. 2003. Mass spectrometry-based proteomics. Nature 422:198–207.
Amin RP, Hamadeh HK, Bushel PR, Bennett L, Afshari CA, Paules RS. 2002. Genomic interrogation of mechanism(s) underlying cellular responses to toxicants. Toxicology 27:555–563.
Ankley GT, Black MC, Garric J, Hutchinson TH, Iguchi TA. 2005 Framework for assessing the hazard of pharmaceutical materials to aquatic species. In: Williams R, editor. Science for assessing the impacts of human pharmaceutical materials on aquatic ecosystems. Pensacola (FL): Society of Environmental Toxicology and Chemistry (SETAC). p. 183–238.

Ankley GT, Daston GP, Degitz SJ, Denslow ND, Hoke RA, Kennedy SW, Miracle AL, Perkins EJ, Snape J, Tillitt DE, et al. 2006. Toxicogenomics in regulatory ecotoxicology. Environ Sci Technol 40:4055–4065.

Ankley GT, Villeneuve DL. 2006. The fathead minnow in aquatic toxicology: past, present and future. Aquat Toxicol. 78:91–102.

Boverhof DR, Zacharewski TR. 2005. Toxicogenomics in risk assessment: applications and needs. Toxicol Sci 89:352–360.

Bundy JG, Spurgeon DJ, Svendsen C, Hankard PK, Weeks JM, Osborn D, Lindon JC, Nicholson JK. 2004. Environmental metabonomics: applying combination biomarker analysis in earthworms at a metal contaminated site. Ecotoxicology 13:797–806.

Bushel PR, Hamadeh HK, Bennett L, Green J, Abelson A, Misener S, Afshari CA, Paules RS. 2002. Computational selection of distinct class- and subclass-specific gene expression signatures. J Biomed Inform 35:160–170.

Colbourne JK, Singan VR, Gilbert DG. 2005. wFleaBase: The *Daphnia* genome database. BMC Bioinformatics 7:45.

Dunn WB, Bailey NJC, Johnson HE. 2005. Measuring the metabolome: current analytical technologies. Analyst 130:606–625.

European Commission. 2001. White paper on the strategy for a future chemicals policy. Document COM 88. Brussels.

Fielden MR, Eynon BP, Natsoulis G, Jarnagin K, Banas D, Kolaja KL. 2005. A gene expression signature that predicts the future onset of drug-induced renal tubular toxicity. Toxicol Pathol 33:675–683.

Fields S. 2001. Proteomics and the future. Science 4:87–102.

Fountoulakis M. 2004. Peptide sequencing by mass spectrometry. Mass Spectrom Rev 23:231–245.

Hamadeh HK, Bushel PR, Jayadev S, DiSorbo O, Bennett L, Li L, Tennant R, Stoll R, Barrett JC, Paules RS, et al. 2002. Prediction of compound signature using high-density gene expression profiling. Toxicol Sci 67:232–240.

Handam M, Righetti P. 2003. Elucidating the cell proteome. Mass Spectrom Rev 22:182–194.

Hess KR, Zhang W, Baggerly KA, Stivers DN, Coombes KR. 2001. Microarrays: handling the deluge of data and extracting reliable information. Trends Biotechnol 19:463–468.

Hutchinson TH, Ankley GT, Segner H, Tyler CR. 2006. Screening and testing for endocrine disruption in fish-biomarkers as signposts not traffic lights in risk assessment. Environ Health Perspect 114(Suppl. 1):106–114.

Ideker T, Thorsson V, Ranish JA, Christmas R, Buhler J, Eng JK, Bumgarner R, Goodlett DR, Aebersold R, Hood L. 2001. Integrated genomic and proteomic analyses of a systematically perturbed metabolic network. Science 292:929–934.

Johnson CD, Balagurunathan Y, Lu KP, Tadesse M, Falalatpisheh MH, Carroll RJ, Dougherty ER, Afshari CA, Ramos KS. 2003. Genomic profiles and predictive biological networks in oxidant-induced atherogenesis. Physiol Genomics 13:263–285.

Kimura T, Jindo T, Narita T, Naruse K, Kobayashi D, Shin-I T, Kitagawa T, Sakaguchi T, Mitani H, Shima A, et al. 2004. Large-scale isolation of ESTs from medaka embryos and its application to medaka developmental genetics. Mech Dev 121:915–932.

Larkin P, Folmar LC, Hemmer MJ, Poston AJ, Denslow ND. 2003. Expression profiling of estrogenic compounds using a sheepshead minnow cDNA macroarray. EHP Toxicogenom 111:29–36.

Lettieri T. 2006. Recent applications of DNA microarray technology to toxicology and ecotoxicology. Environ Health Perspect 114:4–9.

Lühe A, Suter L, Ruepp S, Singer T, Weiser T, Albertini S. 2005. Toxicogenomics in the pharmaceutical industry: hollow promises or real benefit? Mutat Res 575:102–115.

MacGregor JT. 2003. The future of regulatory toxicology: impact of the biotechnology revolution. Toxicol Sci 75:236–248.

Merrick BA, Bruno ME. 2004. Genomic and proteomic profiling for biomarkers and signature profiles of toxicity. Curr Opin Mol Ther 6:600–607.

Miracle AL, Toth G, Lattier DL. 2003. The path from molecular indicators of exposure to describing dynamic biological systems in an aquatic organism: microarrays and the fathead minnow. Ecotoxicology 12:457–462.

Naciff JM, Jump ML, Torontali SM, Carr GJ, Tiesman JP, Overmann GJ, Daston GP. 2002. Gene expression profile induced by 17α-ethynylestradiol, bisphenol A, and genistein in the developing female reproductive system of the rat. Toxicol Sci 68:184–199.

Naruse K, Hori H, Shimizu N, Kohara Y, Takeda H. 2004. Medaka genomics: a bridge between mutant phenotype and gene function. Mech Dev 121:619–618.

Nicholson JK, Connelly J, Lindon JC, Holmes E. 2002. Metabolomics: a platform for studying drug toxicity and gene function. Nat Rev 1:153–162.

Nuwaysir EF, Bittner M, Trent J, Barrett JC, Afshari CA. 1999. Microarrays and toxicology: the advent of toxicogenomics. Mol Carinog 24:153–159.

Pennington S, Wilkins M, Dunn M. 1997. Proteome analysis: from protein characterization to biological function. Trends Cell Biol 7:168–173.

Rasooly RS, Henken D, Freeman H, Tompkins L, Badman D, Briggs J, et al. 2003. Genetic and genomic tools for zebrafish research: the NIH zebrafish initiative. Dev Dyn 228:490–496.

Robertson DG. 2005. Metabonomics in toxicology: a review. Toxicol Sci 85:809–822.

Schmidt CW. 2004. Metabolomics: What's happening downstream of DNA. Environ Health Perspect 112:A410–A415.

Shrader EA, Henry TR, Greeley MS Jr, Bradley BP. 2003. Proteomics in zebrafish exposed to endocrine disrupting chemicals. Ecotoxicology 12:485–488.

Silva J, Geromanos S. 2005. Quantitative proteomic analysis by accurate mass RT-pairs. Anal Chem 77:2187–2200.

Snape JR, Maund SJ, Pickford DB, Hutchinson TH. 2004. Ecotoxicogenomics: the challenge of integrating genomics into aquatic and terrestrial ecotoxicology. Aquat Toxicol 67:143–154.

[SETAC] Society of Environmental Toxicology and Chemistry. 1992. Biomarkers—biochemical, physiological, and histological markers of anthropogenic stress. SETAC Special Publications Series. Ann Arbor (MI): Lewis.

Stentiford GD, Viant MR, Ward DG, Johnson PJ, Martin A, Wenbin W, Cooper HJ, Lyons BP, Feist SW. 2005. Liver tumors in wild flatfish: a histopathological, proteomic and metabolomic study. OMICS 9:281–299.

Thorgaard GH, Bailey GS, Williams D, Buhler DR, Kaattari SL, Ristow SS, Hansen JD, Winton JR, Bartholomew JL, Nagler JJ, et al. 2002. Status and opportunities for genomics research with rainbow trout. Comp Biochem Physiol B Biochem Mol Biol 133:609–646.

Tilton SC, Gerwick LG, Hendricks JD, Rosato CS, Corley-Smith G, Givan SA, Bailey GS, Bayne CJ, Williams DE. 2005. Use of a rainbow trout oligonucleotide microarray to determine transcriptional patterns in aflatoxin B_1-induced hepatocellular carcinoma compared to adjacent liver. Toxicol Sci 88:319–330.

van der Ven K, De Wit M, Keil D, Moens L, Van Leemput K, Naudts B, De Coen W. 2005. Development and application of a brain-specific cDNA microarray for effect evaluation of neuro-active pharmaceuticals in zebrafish (*Danio rerio*). Comp Biochem Physiol B Biochem Mol Biol 141:408–417.

Viant MR. 2003. Improved methods for the acquisition and interpretation of NMR metabolomic data. Biochem Biophys Res Commun 310:943–948.

Viant MR, Rosenblum ES, Tieerdema RS. 2003. NMR-based metabolomics: a powerful approach for characterizing the effects of environmental stressors on organism health. Environ Sci Technol 37:4982–4989.

von Schalburg KR, Rise ML, Cooper GA, Brown GD, Gibbs AR, Nelson CC, Davison WS, Koop BF. 2005. Fish and chips: various methodologies demonstrate utility of a 16 000-gene salmonid microarray. BMC Genomics 6:126.

Waters MD, Fostel JM. 2004. Toxicogenomics and systems toxicology: aims and prospects. Nat Rev Genet 5:936–948.

Williams TD, Gensberg K, Minchin SD, Chipman JK. 2003. A DNA expression array to detect toxic stress response in European flounder (*Platichthys flesus*). Aquat Toxicol 65:141–157.

[WHO] World Health Organization. 2002. ICPS global assessment of the state of the science of endocrine disruptors. WHO/PCS/EDC/02.2. Geneva: International Program on Chemical Safety.

Yu U, Lee SH, Kim YJ, Kim S. 2004. Bioinformatics in the post-genome era. J Biochem Mol Biol 37:75–82.

2 Application of Genomics to Screening

*Sean W Kennedy, Susan Euling, Duane B Huggett,
L Wim De Coen, Jason R Snape,
Timothy R Zacharewski, and Jun Kanno*

CONTENTS

2.1 Introduction 14
2.2 Currently Used Screening Assays 15
    2.2.1 Single Endpoint Assays 15
    2.2.2 Multiple Endpoint Assays 16
2.3 The Promise of Genomics for Screening Assays 17
    2.3.1 International Efforts by Government and Industry 17
    2.3.2 Lessons from Human Health Assessments 18
    2.3.3 Potential Use of Genomics in Ecological Risk Assessment 19
    2.3.4 Specific Examples of Applications of Genomics in
         Ecotoxicology 20
2.4. Current Limitations of Genomic Technologies in Ecological Screening 20
    2.4.1. Variability and Reproducibility 20
    2.4.2 Biological Challenges 21
2.5. Priority Research Questions and Recommendations 21
    2.5.1 Which Currently Available Genomic Techniques Are Most
         Reliable and Appropriate for Application to Ecotoxicology
         Screens? 22
    2.5.2 What Is the Effect of Different Conditions and Different Life
         Stages of Exposure on Gene Expression for a Given Species? 24
    2.5.3 Genetic Sequence Data for New Ecological Species: Which
         Species Have Greatest Maturity in Terms of Current Data and
         Resources? 24
    2.5.4 Which Genomics Techniques Are Expected to Be Ready for
         Ecotoxicology Screens in the Near Future (5 to 10 Years)? 25
    2.5.5 What Is the Best Approach to Understand Cross-Species
         Genomic Similarities and Differences in Response That
         Are Predictive of Cross-Species in Vivo Endpoint Response
         Differences? 26
2.6. Conclusions and General Recommendations 27
References 28

2.1 INTRODUCTION

Screening assays are used in ecotoxicology to estimate hazard, determine exposure, prioritize higher tiered testing or identify specific modes of action (MOAs) of chemicals. Screening assays are designed to be 1) sensitive for detecting the hazard or exposure of concern (i.e., designed to err on the side of false positives over false negatives), 2) relatively rapid, and 3) relatively inexpensive. Although screening assays do not necessarily need to be highly specific in nature, it is preferable when they identify the MOA of the chemicals of interest. Many screening assays are conducted in vitro rather than in vivo to allow for a rapid, inexpensive, and more directed screen. In recent years, international pressures to reduce testing costs and use fewer animals have increased substantially, which encourages the application of existing in vitro assays and the development of new screening assays. In vitro screening assays can be more sensitive and specific than in vivo assays, which often determine apical endpoints such as survival, growth, and reproduction. Since an effective screening assay errs on the side of false positives, chemicals determined to have no deleterious effect in an effective in vitro screen should not have adverse effects in an in vivo test.

Genomic technologies offer promise for identifying and characterizing the biological activity of chemicals, including drugs, biocides, pesticides, and other classes of industrial compounds found in the environment. The emergence of toxicogenomics provides the ability to comprehensively determine changes in gene, protein, and metabolite expression within an organism, tissue, or cell exposed to a toxicant; thus, they could play an important role in screening and hazard assessments for human health and ecological risk assessments. Toxicogenomics, as defined in this book, is the study of the global response of a genome to a chemical, at the levels of mRNA expression (transcriptomics), protein expression (proteomics), and metabolite profiling (metabolomics).

Studies in simpler organisms such as yeast, fly, and worm demonstrate that individual responses are not independent, but form a network of regulation (Giot et al. 2003; Babu et al. 2004; Li et al. 2004; Tong et al. 2004). Therefore, it is likely that more predictive screens will utilize the simultaneous measurement of patterns of expression from multiple genes (mRNA, proteins and metabolites) as opposed to a single response (e.g., cytochrome P4501A [CYP1A] for dioxins and dioxin-like compounds, and cholinesterase for organophosphate and carbamate pesticides). Such global gene expression screens would help minimize the potential for false negatives and increase confidence in their utility for risk assessments.

Challenges for applying genomic technologies as screens to support decision making in ecotoxicology (e.g., prioritization for higher tiered testing, regulations) will include the development, rigorous validation, and acceptance of the utility of new methods for regulatory purposes. Validation procedures are particularly rigorous in cases when an in vitro assay is being developed as a potential replacement of an internationally recognized and validated in vivo assay (http://iccvam.niehs.nih.gov/; http://ecvam.jrc.it/index.htm). Acceptance of genomic-based assays will require collaborative efforts by research and regulatory scientists, and ongoing collaborative research is currently providing focus and information for national and international validation efforts.

The purpose of this chapter is to describe some of the current and planned applications of genomic technologies to screening assays for use in ecotoxicology decision making, including risk assessment. This review focuses on defining the limitations and research needs in the area of genomic technologies for screening; this is not a comprehensive review of genomic technologies or screening tools. Since there are no truly global genomic screens currently used for regulatory purposes, the discussion of current and planned genomic technologies also includes single protein and mRNA assays to place future plans into context within the continuum of current capabilities. Herein, the discussion of future uses of genomics in screening is focused on mRNA expression technologies because they are currently more advanced in their development than either proteomic or metabolomic technologies. However, we anticipate that proteomic, metabolomic, and other genomic technologies will eventually be developed into very useful screening assays.

2.2 CURRENTLY USED SCREENING ASSAYS

2.2.1 SINGLE ENDPOINT ASSAYS

It is useful to review briefly a few key characteristics of biochemical, molecular, and in vitro assays currently used for ecotoxicological decision making because existing assays can be considered as precursors to emerging genomic-based assays. First, many existing bioassays do not predict toxicity directly, but instead indicate that a particular biochemical pathway has been impacted after exposure to the chemical. Second, existing screening assays can be used to assess the impact after exposure to a single chemical or mixtures of chemicals. For example, there are bioassays that are useful for measuring the potencies of specific dioxin-like chemicals or complex mixtures of dioxin-like chemicals. Third, there are bioassays that can be used to predict differences in species sensitivity to certain classes of chemicals. For example, avian species differ substantially in sensitivity to dioxins and other aryl hydrocarbon receptor (AHR) agonists, and in vitro screening assays have been developed to predict such differences (Kennedy et al. 1996; Karchner et al. 2006). Fourth, most, if not all, biochemical or molecular screening assays can be described as "supervised." By definition, the endpoints of supervised assays are specifically selected (e.g., CYP1A, vitellogenin) and most supervised assays measure only one endpoint or a small number of endpoints.

Within a regulatory context, some of the most widely utilized gene and protein expression screens for ecotoxicology decision-making purposes are related to the endocrine system. For example, vitellogenin (VTG) mRNA and protein expression as downstream indicators of estrogen or androgen exposure were developed to help screen for chemicals that might affect oviparous vertebrates. Vitellogenin is an egg-yolk precursor synthesized in the liver whose formation is principally linked to the stimulation of the estrogen receptor, although recent studies have indicated that VTG production may also be responsive to androgenic compounds (Ankley et al. 2005). The measurement of VTG mRNA or protein can be accomplished under a variety of experimental conditions including isolated fish hepatocytes, liver slices, 14 to 21 d in vivo fish screens (Ankley et al. 2001; Schmieder et al. 2004; Olsen et al. 2005), and avian cell cultures (Lorenzen et al. 2003).

Most current screening assays lack the ability to consider multiple MOAs that may lead to an adverse effect and thus potential effects of chemicals on organisms could be overlooked. For example, screening a sample for estrogenic activity using a receptor-binding assay may yield a negative result, but the sample could contain high levels of mutagens. Even if 2 or 3 assays are available, one could come to incorrect conclusions. For example, assays available for estimating the potential toxic effects of a compound might determine no mutagenic effects, no vitellogenin induction, and no CYP1A induction. However, the compound could elicit a toxic response to some organisms due to other MOAs, such as an adverse effect on steroidogenesis. Moreover, many compounds must first be metabolized before they elicit adverse effects. In summary, most (if not all) existing screening assays provide only a relatively small description of effects that may occur in whole organisms.

2.2.2 Multiple Endpoint Assays

In this chapter, we advocate the development of screening assays that would use genomic technologies to develop multiple endpoint assays. Multiple endpoint assays are useful because toxic responses to chemicals are usually the consequence of complex physiological responses affected by a number of factors including the timing and duration of exposure, environment, and genetics. It is likely that an individual chemical operates via more than one MOA, and different MOAs might depend upon the duration of exposure, developmental stage, target tissue, and dose. For example, a polycyclic aromatic hydrocarbon (PAH) might elicit adverse effects through multiple mechanisms, including those associated with binding to globulins, interactions with receptors, bioactivation to compounds that form DNA adducts, generation of reactive oxygen species, and/or inhibition of steroidogenesis.

Genomic technologies have the potential to detect global effects of a chemical on mRNA transcription, protein expression, and metabolite formation and provide more detailed mechanistic information than traditional screening assays are capable of providing. In conjunction with in vivo toxicology, genomic technologies could also be useful for developing screening tools that are more informative of MOAs and differences in sensitivity among species, strains, and individuals.

There is an opportunity to develop suites of mechanistically based biomarkers using genomic technologies that are predictive of toxicity, while reducing uncertainties in extrapolations across species. For example, a more comprehensive evaluation of effluent estrogenicity across species might be possible with emerging genomic technologies. In addition, toxicity identification evaluations (TIE) are often performed to identify which fractions contain compounds that elicit particular biochemical effects (Ankley and Schubauer-Berigan 1995). Screening with genomic approaches may better direct TIE procedures since a broader suite of endpoints could be measured.

Organisms are rarely exposed to a single chemical of potential concern, and relatively subtle toxic responses such as compromised reproductive capacity or immunotoxic effects can result from exposure to mixtures of chemicals. Some mixtures contain chemicals that singly may not cause adverse effects at particular concentrations but that, when present as complex mixtures, can cause unpredictable effects.

Application of Genomics to Screening

Genomic-based screening methods could identify the potential effects of complex mixtures by evaluating chemical signatures (Hamadeh et al. 2002).

Two examples of multiple endpoint assays that show promise for providing information valuable for aiding regulatory decisions are assays that measure steroidogenesis and stress gene responses.

Steroidogenesis: A steroidogenesis assay that has attracted considerable recent attention is the human adrenocortical carcinoma cell line H295R. These cells express all the key enzymes necessary for steroidogenesis, and quantitative polymerase chain reaction (Q-PCR) assays were developed to determine expression patterns of 11 steroidogenic genes (Zhang et al. 2005). Zhang and colleagues demonstrated that the assay could be used for the development of compound-specific response profiles and for screening for the effects of chemicals on various steps in steroidogenesis. The potential of the cell line for screening effects of model compounds on steroidogenesis has been confirmed by different groups (Heneweer et al. 2005; Muller-Vieira et al. 2005; Gracia et al. 2006; Hecker et al. 2006) and it has recently been applied to screen for steroidogenic effects of contaminants in sediments (Blaha et al. 2006).

Stress gene responses: The transcriptional responses of 14 stress promoter or response element chloramphenicol-acetyl transferase (CAT) fusion constructs that are stably integrated into human carcinoma HepG2 cells show considerable promise as a multiple endpoint assay (Todd et al. 1995). Applications range from testing metals (Mumtaz et al. 2002; Tchounwou et al. 2003) to compounds with military use (Miller et al. 2004) to air pollutants (Vincent et al. 1997; Goegan et al. 1998).

2.3 THE PROMISE OF GENOMICS FOR SCREENING ASSAYS

2.3.1 INTERNATIONAL EFFORTS BY GOVERNMENT AND INDUSTRY

Government departments and agencies worldwide are now expressing interest in the possible application of genomic technologies to the development of new screening methods. In addition to numerous research projects funded by several countries in Asia, Europe, and North America, the Organization for Economic Co-operation and Development (OECD) and the International Program on Chemical Safety (IPCS) of the World Health Organization (WHO) are jointly developing an internationally coordinated program on toxicogenomics for application to human and environmental health assessment.

To help stimulate this work, workshops in Berlin (2002) and Kyoto (2004) were held to develop strategies to explore the future applications of toxicogenomics for human and environmental health assessments (http://www.oecd.org/document/29/0,2340,en_2649_34377_34704669_1_1_1_1,00.html). Both workshops identified the need for a strategic plan to enable genomic-based technologies to be transferred from fundamental research to techniques with clear regulatory applications. As such, an OECD/IPCS scientific advisory board was established, and "proof-of-principle" studies were initiated in 2006. Despite many hurdles to incorporating genomics methods into regulatory frameworks (Balbus 2005; http://www.environmentaldefense.org/documents/4532_ToxicHarnessingPower.pdf), it is our opinion that application of genomics to screening methods is inevitable because of the power of the information gained from performing microarray and other multiple endpoint studies.

Some of the most impressive examples of the development of genomic technologies for use in screening are happening within the pharmaceutical industry. For example, genomic methods are being used to develop individualized therapies with the intention of minimizing risk and maximizing efficacy. As such, the US Food and Drug Administration (FDA) has published a guidance document for submission of pharmacogenomic data(Frueh2006;http://www.fda.gov/OHRMS/DOCKETS/98fr/2003d-0497-gdl0002.pdf) as part of the registration process. Other countries have recently published similar documents (http://www.hc-sc.gc.ca/dhp-mps/alt_formats/hpfb-dgpsa/pdf/brgtherap/draft_pharmaco_ebauche_e.pdf). The FDA has also developed ArrayTrack, a free bioinformatics resource for DNA microarray and systems biology data that facilitates the management, analysis, and interpretation of transcriptomics data within a single package (http://www.fda.gov/nctr/science/centers/toxicoinformatics/ArrayTrack/).

Many of the principles just discussed are relevant to the application of toxicogenomics to ecotoxicology. Criteria for the identification, evaluation, and validation of biomarkers for ecotoxicology and human health have much in common.

2.3.2 Lessons from Human Health Assessments

Some lessons from the use of genomic data for human health assessments may be informative for application of genomics to ecological risk assessments and screening. Genomic technologies have shown considerable promise in various areas of biomedical research. For example, some cancer diagnostic tests use mRNA transcript profiles to characterize tumor properties more accurately (e.g., metastatic vs. primary, responsive vs. nonresponsive), and similar approaches have been used to identify pathogens based on mRNA transcript profiles. Metabolomics has been used to identify subpopulations predisposed to heart disease and diabetes. Moreover, as research advances, the ability to differentiate between adaptive and toxic responses continues to improve.

At least 2 examples of USEPA human health assessments have evaluated toxicogenomic data and used them to inform the mode of action. In the recent USEPA acetochlor risk assessment, mRNA expression data were used to inform the MOA for nasal tumor formation in rats exposed to acetochlor (USEPA 2004; Clayson 1994; Genter et al. 2002). Due to cross-species MOA information, it was argued that this MOA was not relevant to human cancer risk assessment. The external peer review draft of the Integrated Risk Information System (IRIS) dibutyl phthalate (DBP) health assessment is another example where microarray and Q PCR data were used in a weight-of-evidence approach (http://cfpub.epa.gov/ncea/cfm/recordisplay.cfm?deid=155707) to corroborate the MOA and selection of the critical effect of a reduction in fetal testicular testosterone. After in utero DBP exposure in the rat, expression of a number of genes involved in steroidogenesis was significantly altered in the fetal testis (Shultz et al. 2001; Liu et al. 2005).

2.3.3 POTENTIAL USE OF GENOMICS IN ECOLOGICAL RISK ASSESSMENT

While human health risk assessment is concerned with risk of chemicals to individuals of one species, ecological risk assessment (ERA) is usually primarily concerned with the effect of a chemical on populations, communities, and ecosystems encompassing hundreds of species. As such, an ERA attempts to find effects of chemicals on survival, growth, development, and reproduction of organisms. The endpoints for ERA screening assays are selected by various criteria such as the intended use of the chemical, the anticipated route and duration of exposure, and the bioavailability of the chemical.

Considering the multitude of chemicals and situations that need to be considered for ecological testing, it seems logical that genomic technologies could play a role in screening and prioritizing within a regulatory ERA framework. From a prospective ERA stance, the registration of new chemicals can sometimes be best described as a "shot in the dark". Except for pharmaceuticals and pesticides where the MOA is usually known in vertebrates, there are no or limited background data for most chemicals in commerce that can help guide testing. Therefore, they are screened using quantitative structure–activity relationships (QSAR) or a generic base set of laboratory tests (e.g., *Daphnia* sp., algae, fish acute tests). These QSARs or generic tests can provide some toxicity and mechanistic information, but may not necessarily be predictive of toxicity for chemicals with unknown MOAs. Further, in cases when one MOA is identified, the chemical may act via a secondary MOA that could produce more detrimental effects. For instance, the commonly used antibacterial agent triclosan was recently identified as a modulator of the thyroid system in *Rana catesbeiana* (Veldhoen et al. 2006). Previous studies using standardized bioassays were unable to categorize this compound as potentially impacting the thyroid axis (Orvos et al. 2002).

Genomics has the advantage of identifying affected pathways as opposed to the expression of single genes. Therefore, genomic technologies could assist in the identification of potentially impacted pathways in both retrospective and prospective chemical risk assessment. For retrospective risk assessments, often one has little data regarding the cause of a given impact in an environment (e.g., USEPA Superfund sites). In this case, genomic technologies could be used to determine the altered biochemical and physiological processes. With this type of information, it may be possible to identify the causative agents. Historically, single gene or protein measures have been extremely useful in determining when a given impact was due, for example, to a metal or certain classes of organic substances using metallothionein or cytochrome P4501A (Huggett et al. 1992; Peakall 1992). While these types of single gene or protein measures can be extremely useful, they can be cumbersome and costly to conduct for identifying every potential MOA of a contaminant at a field site. A more global genomic approach would provide a more integrative data set that could be used for the environmental risk assessment since it provides patterns of gene expression and affected processes and pathways.

2.3.4 Specific Examples of Applications of Genomics in Ecotoxicology

The number of studies using ecotoxicogenomics is increasing. Next, we provide examples of 3 studies of relevance to this chapter because the data obtained can potentially be useful for screening for potential harmful effects of chemicals on organisms.

Larkin, Knoebl, and Denslow (2003) examined the effects of exposure to 2 estrogenic chemicals on mRNA expression level in 2 different species of fish: the sheepshead minnows (Larkin, Folmar, et al. 2002; Larkin, Folmar, et al. 2003) and the largemouth bass (Larkin, Sabo-Attwood, et al. 2002). A cDNA array derived from genes isolated by differential display RT-PCR was used to examine mRNA expression profiles in the livers of the fish. Their findings showed that their cDNA array was sensitive enough to detect changes in mRNA expression at environmentally relevant concentrations of a contaminant. Furthermore, the study demonstrated that 2 estrogen receptor agonists—17β-estradiol and 4-nonylphenol—had similar, but not identical, genomic signatures.

In another example of the use of genomics in ecotoxicology, van der Ven and colleagues (2006) studied effects of the neuropharmaceutical, mianserin, on zebrafish (*Danio rerio*) using a brain-specific custom cDNA microarray. After aquatic exposure of male and female zebrafish to mianserin at 3 concentrations for 3 time points, RNA was extracted from brain tissue and used for microarray hybridization. In parallel, the impact of exposure to mianserin on egg production, fertilization, and hatching success was determined. Associations between expression of important neuroendocrine-related genes and egg viability were determined. The authors concluded that the integration of reproductive effects with mRNA expression data in the brain of zebrafish will be a useful approach for risk prioritization and environmental research on neuropharmaceuticals.

A third interesting example is the work that Poynton and colleagues (2007) carried out with *Daphnia magna*. Using a custom *D. magna* cDNA microarrray, they identified distinct expression profiles in response to exposure to sublethal concentrations of copper, cadmium, and zinc; they discovered specific biomarkers of exposure, including 2 probable metallothioneins and a ferritin mRNA. The mRNA express patterns supported known MOAs of metal toxicity and, most importantly, revealed novel MOAs, including zinc inhibition of chitinase activity.

2.4. CURRENT LIMITATIONS OF GENOMIC TECHNOLOGIES IN ECOLOGICAL SCREENING

2.4.1. Variability and Reproducibility

Initial reports of DNA microarray data variability and the lack of repeatability have been gradually replaced by more positive and reliable findings in recent years. For example, Larkin and colleagues (2005) showed that with a careful study design and a clear hypothesis, 2 independent platforms (Affymetrix gene chips and cDNA arrays) gave similar results. In another study, Irizarry and colleagues (2005) made a comparison over platforms and laboratories (Affymetrix genechips, spotted cDNA and oligonucleotide arrays). Variation among the different laboratories was substantial, even with the same platform. However, the results from the laboratories with the most reproducible data corresponded quite well. This comparison showed that the differences are not intrinsically linked to the platform; rather, the quality control of the methodologies is more likely to impact the quality of the outcome of array experiments. Bammler and colleagues (2005) came to the same conclusion: Standardizing protocols across laboratories and platforms greatly increased reproducibility in an interplatform and interlaboratory comparison.

The 3 studies cited previously demonstrate that the degree of expertise in performing a microarray experiment is the factor that most likely influences the quality of microarray results. However, differences between various platforms may be linked to the intrinsic properties of the probes and platform limitations (e.g., cDNA vs. oligonucleotide).

2.4.2 Biological Challenges

The most obvious biological challenge is the large diversity of species that one might consider using for screening assays and the limited amount of genetic information that exists for most species of potential interest. Comparative genomic approaches can be powerful for identifying conserved responses across species, but there are significant challenges to such approaches. Because gene annotation for even the best characterized species (human and mouse) is incomplete and limited to similarity comparisons determined by computational methods, comprehensive interpretation of large genomic data sets is difficult. In addition, the lack of sequence information for wildlife species limits the ability to identify orthologs (i.e., genes in different species that are similar to each other and originated from a common ancestor, regardless of their function).

While the limited amount of DNA sequence information for wildlife species currently compromises the ability to elucidate mechanisms and biochemical pathways associated with toxicity, DNA sequencing costs are expected to decrease substantially in the future. Lower sequencing costs will therefore alleviate some of the current problems associated with the limited amount of genetic information for wildlife species. Moreover, as efforts become less directed to whole genome sequencing and more to functional annotation of the genome, resources to elucidate mechanisms will improve.

2.5. PRIORITY RESEARCH QUESTIONS AND RECOMMENDATIONS

Research recommendations for the application of genomics to ecotoxicology screening include both general technological needs (e.g., improving the reproducibility of microarray studies) and specific tools and baseline data for species relevant to ecotoxicology testing and monitoring. In this chapter our recommendations are restricted to research needs specific to use of genomics in screening assays. In order to use genomic technologies for ecotoxicological decision making, screening assays need to be developed, standardized, and validated for species relevant to ERAs. Research needs are organized under 5 priority research questions identified by our workgroup (Table 2.1).

TABLE 2.1
Priority Research Questions and Recommendations for the Development of Genomics-Based Ecotoxicological Screening Assays

Research Question[a]	Research Recommendation
1. Which genomics techniques currently are the most reliable and appropriate for application to ecotox screens?	Validation of transcriptomics techniques for use in ecotox screening
2. What is the effect of different conditions and different life stages of exposure on gene expression for a given species?	Perform microarray studies to address the following questions: What are the impacts of pH, salinity, and temperature? What is the expression profile over developmental time? What is the inherent intraspecies variability in response (not related to technical issues)?
3. Genetic sequence data for new ecological species: Which species have the greatest maturity in terms of current data and resources?	Gene annotation for new but mature ecological species. Development of improved "open" methods of gene expression that do not require sequence data from a species (e.g., SAGE techniques, differential display)
4. Which genomics techniques are expected to be ready for ecotoxicology screens in the near future (5 to 10 years)?	Studies to validate proteomics. Studies to validate metabolomics
5. What is the best approach to understand cross-species genomic similarities and differences in response that are predictive of cross-species in vivo endpoint response differences?	Chemical- and species-specific case studies to determine key gene expression changes linked to adverse outcomes. Development of strategies to use gene expression data quantitatively (i.e., determination of the level of change that will lead to an adverse outcome)

[a] These are not listed in order of priority.

2.5.1 WHICH CURRENTLY AVAILABLE GENOMIC TECHNIQUES ARE MOST RELIABLE AND APPROPRIATE FOR APPLICATION TO ECOTOXICOLOGY SCREENS?

Transcriptomic methods, particularly cDNA microarrrays, are currently further advanced and appropriate for screening assays than proteomic and metabolomic methods. While microarrays are considerably more reliable than they were a few years ago, candidate genes or suites of genes that would be used for a proposed screening method should be validated using another technique, such as Q-PCR. Species that have well-characterized genomes deserve attention as candidates for screening assays even if they are not considered the most ecologically relevant species (e.g., zebrafish).

Validation of transcriptomic data is particularly critical, given 1) the maturity of the science, 2) the volume and complexity of the information generated, 3) the number of variables that can introduce uncertainty, 4) the number of platforms, 5) the variety of bioinformatic techniques for data analysis, and 6) the rate of technological innovation. Information required to assist with the validation of assays for use in

TABLE 2.2
Information Required to Assist in Validation of Assays for Use in Regulatory Decision Making[a]

System Definition	Data Requirements
Test or assay definition	Test protocol including standard operating procedure (SOP). Scientific purpose of the test. Mechanistic basis of the test endpoint, if known. Training set of chemicals including chemical structures and the experimental data. Details of data analysis and interpretation. Provisional domain of applicability
Intralaboratory variability	Assessment of reproducibility within the same laboratory to include operator and temporal variability
Transferability	Assessment of the ease of the transferability and accessibility to the assay being proposed. Preliminary assessment of interlaboratory reproducibility in a second laboratory
Interlaboratory variability	Ring test to assess reproducibility in 2 to 4 laboratories
Predictive capability	Assessment of the predictive capacity of the assay through the use of chemicals not used in the development of the assay
Application domain	Definition of the chemical classes and or ranges of test assay or method endpoints for which reliable predictions can be made
Performance standards and validity criteria	Definition of positive and negative reference substances that can be used to demonstrate the equivalence in performance between a new test and a previously validated test

[a] Adapted from Harting T. et al. 2004. ATLA. 32:467–472.

regulatory decision making are summarized in Table 2.2. The issues of intra- and interlaboratory variability for microarray data have been outlined in the published literature (Shi et al. 2006; Yauk and Berndt 2007).

The European Center for the Validation of Alternative Methods (ECVAM) and the US Interagency Coordinating Committee on the Validation of Alternative Methods (ICCVAM) work to ensure that new and alternative test methods are validated prior to their regulatory use. They have recently published the findings of a workshop on the validation of toxicogenomics-based test methods (Corvi et al. 2006). Three specific focus areas were identified: 1) biological validation of toxicogenomics-based test methods for regulatory decision making, 2) technical and bioinformatics issues related to validation, and 3) validation issues as they relate to regulatory acceptance and utilization of toxicogenomics-based test methods. Corvi et al. (2006) identified 2 biological strategies intended to support regulatory decision making in toxicology. One uses phenotypic anchoring of gene expression changes to identify molecular mechanisms and candidate biomarkers of toxicity (i.e., single genes, proteins, or biological pathways). A second strategy uses predictive gene expression signatures of toxicity. For ecotoxicology, the added dimension of intra- and interspecies susceptibility needs to be considered in order to improve the scientific basis for extrapolation.

2.5.2 What Is the Effect of Different Conditions and Different Life Stages of Exposure on Gene Expression for a Given Species?

One of the main areas of uncertainty in ecotoxicology is extrapolation of toxicity data from laboratory testing to ecosystem effects. Although laboratory testing is essential for identifying mechanisms of toxicity in a standardized and controlled fashion, considerable confusion about the validity of laboratory-to-field extrapolations of toxicity test data remains (Heugens et al. 2001; Selck et al. 2002). Not only does field-based in situ biomonitoring capture possible effects of a myriad of possibly harmful compounds, each of which can elicit a direct toxic effect, but ecological factors, such as food availability, trophic dynamics, seasonality and species interactions also can modulate the effects (Heugens et al. 2001; Fleeger et al. 2003). In these cases, models are needed to describe and understand the interrelationships among the various biotic and abiotic variables that could affect the outcome of an environmental hazard or risk assessment.

Development of tests that expose animals at the most sensitive or critical life stages will improve screening methods. Many ecotoxicogenomic studies have focused on adult- or mixed-stage populations, which may or may not be the most sensitive to contaminant exposure. Attention to the timing of exposure and assessment may also contribute to reducing the variability in expression response; that is, mixed populations will respond differently to certain toxic agents, improving the specificity of the technology.

As the genomic techniques become less variable and less expensive, inclusion of multiple life stages of both exposure and assessment in the experimental designs will provide a rich area for future research. For example, careful design of experiments with organisms at different life stages will allow for identification of critical windows of exposure (e.g., highly sensitive developmental times) and aid in developing expression patterns that may correspond temporally to adverse outcomes. In addition, knowledge about expression profiles during different life stages is extremely important when considering screening within a field context. If expression profiles change over the life span of an organism, then making comparisons between the same age organisms is important. Also, understanding variability of profiles under the normal range of environmental and experimental conditions (e.g., salinity, pH, temperature) must be considered when developing these technologies.

2.5.3 Genetic Sequence Data for New Ecological Species: Which Species Have Greatest Maturity in Terms of Current Data and Resources?

Obtaining genomic data for ecological species that currently lack this information is a necessary first step before genomics screening assays can be developed. We recommend genetic characterization of keystone species from different trophic levels using single nucleotide polymorphisms (SNPs) and other genetic markers. There is a need for gene annotation for mammalian and nonmammalian species. Genomic data from freshwater, marine, and terrestrial species that are relevant to ecological risk assessments are needed. When available, gene annotation information can be made publicly available by deposition to a repository such as the Comparative Toxicogenomics database (Mattingly et al. 2004).

Alternative approaches to further genetic annotation have developed subtractive hybridization techniques to derive targeted cDNA, and such methods are beginning to show considerable promise (van der Ven et al. 2006). Open methods of gene expression such as differential display PCR and serial analysis of gene expression (SAGE) can also be of value for discovering genes affected by toxicants and might be incorporated into a screening method.

2.5.4 Which Genomics Techniques Are Expected to Be Ready for Ecotoxicology Screens in the Near Future (5 to 10 Years)?

Genomics could be used to develop screening methods to classify chemicals into categories based upon similarities in genomic profile. As such, developing customized libraries representing different chemical classes would be useful. For example, subsets of genes whose expression was altered by exposure to estrogenic chemicals (e.g., genistein, bisphenol A, or 17α-ethynylestradiol) have been identified (Naciff, Hess, et al. 2005; Moens et al. 2006). Further work to elucidate whether these genes are linked to an estrogen-related adverse outcome would help validate a subset or all of these genes for a screening assay.

Another possible use of screening assays is the prediction of adverse chronic outcomes. Studies could be carried out to define relationships between early phenotypic anchoring gene expression data and chronic endpoints. In these instances it may be necessary to assess correlations between genomic markers and chronic endpoints using knockout or knockdown studies to establish confirmatory, mechanistic links; this could be difficult for many ecologically relevant species, since no molecular tools to achieve gene-targeted confirmation have been developed.

In mammalian toxicology, first-generation microarrays and microarray protocols were more qualitative than quantitative. Current second-generation microarrays, especially for species for which the whole genome is sequenced, are more quantitative and hence more applicable to dose–response studies. These new arrays have better linearity and reproducibility, including better control over cross-hybridization due to the improved manufacturing strategies and protocols for sample treatment (e.g., use of consistent processing machines for minimization of user-dependent variability, at least for certain critical processes). Some of these systems are capable of generating quantitative data comparable to Q-PCR data. An interesting new protocol that normalizes microarray data against genomic DNA in order to generate profile-independent data in terms of copy numbers of mRNA on a per-cell basis might be very useful for comparing data among different species (Kanno et al. 2006).

We recommend the development of suites of assays that, as a whole, would provide mechanistic information as well as confidence about predicting an adverse outcome of interest. Effects on mortality, growth, and reproductive success (i.e., number of offspring, malformations, sex ratio) or the outcome of interest should be monitored in addition to genomic measurements. While genomics-based assays currently do not suffice as sole drivers in a screening-based decision-making process and genomic endpoints need to be complemented by at least some higher level endpoint, in the future (5 to 10 years), it is likely that genomic changes linked to adverse outcomes could become reliable screening assays used in decision making.

2.5.5 What Is the Best Approach to Understand Cross-Species Genomic Similarities and Differences in Response That Are Predictive of Cross-Species in Vivo Endpoint Response Differences?

One priority research need is the development of a cross-species genomics database. Such a database would allow for an understanding of the full spectrum of similarities and differences in response across species and, thus, interspecies interpolation would become possible. Using the example of screening for estrogenic compounds (Naciff, Richardson, et al. 2005), development of MOA-based microarray expression signatures across species would be very useful. One could compare the genes whose expression is significantly altered after estrogenic exposure across species. A subset of genes altered across part of the phylogenetic tree would be useful for both hazard identification and understanding the phylogenetic limits to which compounds with specific MOAs likely pose a risk. Interestingly, exposure to environmental estrogens ethinyl estradiol or bisphenol A led to reproductive and developmental effects consistent with an estrogen agonist MOA in snails—species in which estrogen receptors have not been identified (Oehlmann et al. 2000; Jobling et al. 2004; Oehlmann et al. 2006). Therefore, it would be interesting to compare toxicity signatures after exposure to invertebrate versus vertebrate species, for example. Commonalities may allow for the development of one screen across animal species; conversely, differences may allow for the development of multiple screens across animal species.

Since mechanisms required for homeostasis and reproduction are often well conserved across species, comprehensive data-rich approaches can be used to identify conserved mechanisms important in these fundamental processes. Moreover, these mechanisms are also likely targets for disruption by toxic agents when comparable toxicities are conserved across species. Therefore, systematic comparative genomic analyses that comprehensively examine the impact of a chemical agent could identify mechanisms and specific targets important in eliciting toxicity. In order to associate genomic changes with endpoints, these studies would include dose–response and time-course studies, as well as integrate phenotypically anchored endpoints (e.g., histopathology, morphometry, tissue weight, tissue function).

Comparative genomics approaches could provide data useful for the integration of human and ecological risk assessments (Suter et al. 2005). For example, comparative genomic approaches have been used to identify functionally related genes across divergent species (e.g., extrapolations among human, worm, fly, and yeast (Stuart et al. 2003). The completion of genome sequencing of a number of species, including human, mouse, rat, zebrafish, chicken, and other species, provides the resources necessary to identify putative conserved mechanisms of toxicity, at least at the gene expression level. Parallel studies could be performed to identify conserved responses at various levels of biological organization in different sequenced species (e.g., mouse vs. rat vs. zebrafish vs. chicken) using comparable treatment regimens. Through comparative time course and dose–response studies, as well as structure activity relationship (SAR) studies, conserved responses across

Application of Genomics to Screening

different taxa could be identified based on global gene expression patterns (including dose–response relationship and temporal expression) as well as on predicted amino acid sequence homologies. Together, these studies could identify responses based not only on sequence homology but also on conserved temporal and dose–response profile characteristics.

2.6. CONCLUSIONS AND GENERAL RECOMMENDATIONS

Over the past several years, the field of genomics has grown considerably. The use of global gene expression assessment (e.g., microarrays) has slowly begun to replace single gene and protein measures. This enables measurement of complex patterns of gene expression responses within an organism. Evaluating multiple endpoints via a screening battery has been successful for prioritization and screening purposes. Thus, we propose that efforts to develop screening assays for use in ecological risk assessment should focus on the development of a suite of endpoints that incorporate genomic assays as part of the battery.

While both in vivo and in vitro screens are important, in vitro screens deserve extra attention because they potentially reduce animal use. In our opinion, the need for whole-animal testing will not be eliminated in the near future, but in vitro screens do have great potential to reduce the numbers of organisms used for screening for harmful effects of chemicals.

In order to implement screening technologies, they must be fully validated within the context of the reference compounds essential for a "proof of concept" to allow these technologies to be validated and used within a regulatory context. Reference compound data could be used for investigating across genomic methods (transcriptomics, proteomics, metabolomics), platforms, and tiered testing levels (acute, chronic, field). In considering selection criteria for reference compounds, traditional environmental toxicants (e.g., PCBs, DDT, cadmium) may be excellent candidates as a considerable amount of higher tiered effects data are already available, as well as pharmaceutical materials, owing to the extensive amount of pharmacological and toxicological data already available, including, potentially, genomic information. For instance, Hamadeh and colleagues (2002) have published extensive microarray studies on peroxisome proliferators, such as gemfibrozil and clofibrate, identifying a unique signature profile for this class of compounds. By using pharmaceuticals, cross-species extrapolation information is available for addressing and prioritizing emerging contaminant issues.

It is unlikely that universal screening assays (i.e., assays for detecting every single adverse effect or mechanism) can be expected in the near future. This would involve enormous financial and logistical efforts. However, research is recommended that will pave the way to selected screening assay sets that measure effects that we, as a scientific community, feel are the most pressing. For instance, reproduction and endocrine activity may be viewed as 2 of the most critical current toxicological concerns. Developing screening assays that have been scientifically grounded and validated would greatly aid in the selection of compounds that need further evaluation. The goal of the recommended research is to initiate the development of screening assays using genomic-based tools that are currently or nearly ready for use in ecotoxicology, and provide the first logical step toward using genomic-based screens in ecological risk assessment and other decision making.

REFERENCES

Ankley GT, Jensen KM, Durhan EJ, Makynen EA, Butterworth BC, Kahl MD, Villeneuve DL, Linnum A, Gray LE, Cardon M, et al. 2005. Effects of two fungicides with multiple modes of action on reproductive endocrine function in the fathead minnow (*Pimephales promelas*). Toxicol Sci 86:300–308.

Ankley GT, Jensen KM, Kahl MD, Korte JJ, Makynen EA. 2001. Description and evaluation of a short-term reproduction test with the fathead minnow (*Pimephales promelas*). Environ Toxicol Chem 20:1276–1290.

Ankley GT, Schubauer-Berigan MK. 1995. Background and overview of current sediment toxicity identification evaluation procedures. J Aquatic Ecosyst Stress Recov 4:133–149.

Babu K, Cai Y, Bahri S, Yang X, Chia W. 2004. Roles of bifocal, Homer, and F-actin in anchoring Oskar to the posterior cortex of *Drosophila* oocytes. Genes Dev 18:138–143.

Balbus JM. 2005. Ushering in the new toxicology: toxicogenomics and the public interest. Environ Health Perspect 113:818–822.

Bammler T, Beyer RP, Bhattacharya S, Boorman GA, Boyles A, Bradford BU, Bumgarner RE, Bushel PR, Chaturvedi K, Choi D, et al. 2005. Standardizing global gene expression analysis between laboratories and across platforms. Nat Methods 2:351–356.

Blaha L, Hilscherova K, Mazurova E, Hecker M, Jones PD, Newsted JL, Bradley PW, Gracia T, Duris Z, Horka I, et al. 2006. Alteration of steroidogenesis in H295R cells by organic sediment contaminants and relationships to other endocrine disrupting effects. Environ Int 32:749–757.

Clayson DB, Mehta R, Iverson F. 1994. International commission for protection against environmental mutagens and carcinogens. Oxidative DNA damage—the effects of certain genotoxic and operationally nongenotoxic carginogens. Mutat Res 317:25–42.

Corvi R, Ahr HJ, Albertini S, Blakey DH, Clerici L, Coecke S, Douglas GR, Gribaldo L, Groten JP, Haase B, et al. 2006. Meeting report: Validation of toxicogenomics-based test systems: ECVAM-ICCVAM/NICEATM considerations for regulatory use. Environ Health Perspect 114:420–429.

Fleeger JW, Carman KR, Nisbet RM. 2003. Indirect effects of contaminants in aquatic ecosystems. Sci Total Environ 317:207–233.

Frueh FW. 2006. Impact of microarray data quality on genomic data submissions to the FDA. Nat Biotechnol 24:1105–1107.

Genter MB, Burman DM, Vijayakumar S, Ebert CL, Aronow BJ. 2002. Genomic analysis of alachlor-induced oncogenesis in rat olfactory mucosa. Physiol Genomics 12:35–45.

Giot L, Bader JS, Brouwer C, Chaudhuri A, Kuang B, Li Y, Hao YL, Ooi CE, Godwin B, Vitols E, et al. 2003. A protein interaction map of *Drosophila melanogaster*. Science 302:1727–1736.

Goegan P, Vincent R, Kumarathasan P, Brook J. 1998. Sequential in vitro effects of airborne particles in lung macrophages and reporter CAT-gene cell lines. Toxicol in Vitro 12:25–37.

Gracia T, Hilscherova K, Jones PD, Newsted JL, Zhang X, Hecker M, Higley EB, Sanderson JT, Yu RM, Wu RS, et al. 2006. The H295R system for evaluation of endocrine-disrupting effects. Ecotoxicol Environ Saf 65:293–305.

Hamadeh HK, Bushel PR, Jayadev S, Martin K, DiSorbo O, Sieber S, Bennett L, Tennant R, Stoll R, Barrett JC, et al. 2002. Gene expression analysis reveals chemical-specific profiles. Toxicol Sci 67:219–231.

Hecker M, Newsted JL, Murphy MB, Higley EB, Jones PD, Wu R, Giesy JP. 2006. Human adrenocarcinoma (H295R) cells for rapid in vitro determination of effects on steroidogenesis: Hormone production. Toxicol Appl Pharmacol 217:114–124.

Heneweer M, van den Berg M, de Geest MC, de Jong PC, Bergman A, Sanderson JT. 2005. Inhibition of aromatase activity by methyl sulfonyl PCB metabolites in primary culture of human mammary fibroblasts. Toxicol Appl Pharmacol 202:50–58.

Heugens EH, Hendriks AJ, Dekker T, van Straalen NM, Admiraal W. 2001. A review of the effects of multiple stressors on aquatic organisms and analysis of uncertainty factors for use in risk assessment. Crit Rev Toxicol 31:247–284.

Huggett RJ, Kimerle RA, Mehrle PM Jr, Bergman HL, editors. 1992. Biomarkers: biochemical, physiological, and histological markers of anthropogenic stress. Boca Raton (FL): Lewis Publishers.

Irizarry RA, Warren D, Spencer F, Kim IF, Biswal S, Frank BC, Gabrielson E, Garcia JG, Geoghegan J, Germino G, et al. 2005. Multiple-laboratory comparison of microarray platforms. Nat Methods 2:345–350.

Jobling S, Casey D, Rogers-Gray T, Oehlmann J, Schulte-Oehlmann U, Pawlowski S, Baunbeck T, Turner AP, Tyler CR. 2004. Comparative responses of mollusks and fish to environmental estrogens and an estrogenic effluent. Aquat Toxicol 66(2):207–222.

Kanno J, Aisaki K, Igarashi K, Nakatsu N, Ono A, Kodama Y, Nagao T. 2006. "Per cell" normalization method for mRNA measurement by quantitative PCR and microarrays. BMC Genomics 7:64.

Karchner SI, Franks DG, Kennedy SW, Hahn ME. 2006. The molecular basis for differential dioxin sensitivity in birds: role of the aryl hydrocarbon receptor. PNAS 103:6252–6257.

Kennedy SW, Lorenzen A, Jones SP, Hahn ME, Stegeman JJ. 1996. Cytochrome P4501A induction in avian hepatocyte cultures: a promising approach for predicting the sensitivity of avian species to toxic effects of halogenated aromatic hydrocarbons. Toxicol Appl Pharmacol 141:214–230.

Larkin JE, Frank BC, Gavras H, Sultana R, Quackenbush J. 2005. Independence and reproducibility across microarray platforms. Nat Methods 2:337–344.

Larkin P, Folmar LC, Hemmer MJ, Poston AJ, Denslow ND. 2003. Expression profiling of estrogenic compounds using a sheepshead minnow cDNA macroarray. EHP Toxicogenomics 111:29–36.

Larkin P, Folmar LC, Hemmer MJ, Poston AJ, Lee HS, Denslow ND. 2002. Array technology as a tool to monitor exposure of fish to xenoestrogens. Mar Environ Res 54:395–399.

Larkin P, Knoebl I, Denslow ND. 2003. Differential gene expression analysis in fish exposed to endocrine disrupting compounds. Comp Biochem Physiol B Biochem Mol Biol 136:149–161.

Larkin P, Sabo-Attwood T, Kelso J, Denslow ND. 2002. Gene expression analysis of largemouth bass exposed to estradiol, nonylphenol, and p,p'-DDE. Comp Biochem Physiol B Biochem Mol Biol 133:543–557.

Li Z, Chan C. 2004. Integrating gene expression and metabolic profiles. J Biol Chem. 279:27124–27137.

Liu K, Lehmann KP, Sar M, Young SS, Gaido KW. 2005. Gene expression profiling following in utero exposure to phthalate esters reveals new gene targets in the etiology of testicular dysgenesis. Biol Reprod 73:180–192.

Lorenzen A, Williams KL, Moon TW. 2003. Determination of the estrogenic and antiestrogenic effects of environmental contaminants in chicken embryo hepatocyte cultures by quantitative-polymerase chain reaction. Environ Toxicol Chem 22:2329–2336.

Mattingly CJ, Colby GT, Rosenstein MC, Forrest JN Jr, Boyer JL. 2004. Promoting comparative molecular studies in environmental health research: an overview of the comparative toxicogenomics database (CTD). Pharmacogenomics J 4:5–8.

Miller AC, Brooks K, Smith J, Page N. 2004. Effect of the militarily relevant heavy metals, depleted uranium and heavy metal tungsten-alloy on gene expression in human liver carcinoma cells (HepG2). Mol Cell Biochem 255:247–256.

Moens LN, van der Ven K, van Remortel P, Del Favero J, De Coen WM. 2006. Expression profiling of endocrine-disrupting compounds using a customized *Cyprinus carpio* cDNA microarray. Toxicol Sci 93:298–310.

Muller-Vieira U, Angotti M, Hartmann RW. 2005. The adrenocortical tumor cell line NCI-H295R as an in vitro screening system for the evaluation of CYP11B2 (aldosterone synthase) and CYP11B1 (steroid-11beta-hydroxylase) inhibitors. J Steroid Biochem Mol Biol 96:259–270.

Mumtaz MM, Tully DB, El Masri HA, De Rosa CT. 2002. Gene induction studies and toxicity of chemical mixtures. Environ Health Perspect 110(Suppl 6):947–956.

Naciff JM, Hess KA, Overmann GJ, Torontali SM, Carr GJ, Tiesman JP, Foertsch LM, Richardson BD, Martinez JE, Daston GP. 2005. Gene expression changes induced in the testis by transplacental exposure to high and low doses of 17{alpha}-ethynyl estradiol, genistein, or bisphenol A. Toxicol Sci 86:396–416.

Naciff JM, Richardson BD, Oliver KG, Jump ML, Torontali SM, Juhlin KD, Carr GJ, Paine JR, Tiesman JP, Daston GP. 2005. Design of a microsphere-based high-throughput gene expression assay to determine estrogenic potential. Environ Health Perspect 113:1164–1171.

Oehlmann J, Schulte-Oehlmann U, Tillmann M, et al. 2000. Effects of endocrine disruptors on prosobranch snails (*Mollusca: gastropoda*) in the laboratory. Part I: Bisphenol A and octylphenol as xeno-estrogens. Ecotoxicology 9:383–397.

Oehlmann J, Schulte-Oehlmann U, Bachmann J, Oetken M, Lutz I, Kloas W, Ternes TA. 2006. Bisphenol A induces superfeminization in the ramshorn snail *Marisa cornuarietis* (Gastropoda: Prosobranchia) at environmentally relevant concentrations. Environ Health Perspect 114(Suppl 1):127–133.

Olsen CM, Meussen-Elholm ET, Hongslo JK, Stenersen J, Tollefsen KE. 2005. Estrogenic effects of environmental chemicals: an interspecies comparison. Comp Biochem Physiol C Toxicol Pharmacol 141:267–274.

Orvos DR, Versteeg DJ, Inauen J, Capdevielle M, Rothenstein A, Cunningham V. 2002. Aquatic toxicity of triclosan. Environ Toxicol Chem 21:1338–1349.

Peakall DB. 1992. Animal biomarkers as pollution indicators, Ecotoxicological Series 1. London: Chapman & Hall.

Poynton HC, Varshavsky JR, Chang B, Cavigiolio G, Chan S, Holman PS, Loguinov AV, Bauer DJ, Komachi K, Theil EC, et al. 2007. *Daphnia magna* ecotoxicogenomics provides mechanistic insights into metal toxicity. Environ Sci Technol 41:1044–1050.

Schmieder PK, Tapper MA, Denny JS, Kolanczyk RC, Sheedy BR, Henry TR, Veith GD. 2004. Use of trout liver slices to enhance mechanistic interpretation of estrogen receptor binding for cost-effective prioritization of chemicals within large inventories. Environ Sci Technol 38:6333–6342.

Selck H, Riemann B, Christoffersen K, Forbes VE, Gustavson K, Hansen BW, Jacobsen JA, Kusk OK, Petersen S. 2002. Comparing sensitivity of ecotoxicological effect endpoints between laboratory and field. Ecotoxicol Environ Saf 52:97–112.

Shi L, Reid LH, Jones WD, Shippy R, Warrington JA, Baker SC, Collins PJ, de Longueville F, Kawasaki ES, Lee KY, et al. 2006. The MicroArray Quality Control (MAQC) project shows inter- and intraplatform reproducibility of gene expression measurements. Nat Biotechnol 24:1151–1161.

Shultz VD, Phillips S, Sar M, Foster PM, Gaido KW. 2001. Altered gene profiles in fetal rat testes after in utero exposure to di(n-butyl) phthalate. Toxicol Sci 64:233–242.

Stuart JM, Segal E, Koller D, Kim SK. 2003. A gene-coexpression network for global discovery of conserved genetic modules. Science 302:249–255.

Suter GW, Vermeire T, Munns WR Jr, Sekizawa J. 2005. An integrated framework for health and ecological risk assessment. Toxicol Appl Pharmacol 207:611–616.
Tchounwou PB, Yedjou CG, Dorsey WC. 2003. Arsenic trioxide-induced transcriptional activation of stress genes and expression of related proteins in human liver carcinoma cells (HepG2). Cell Mol Biol (Noisy-le-grand) 49:1071–1079.
Todd MD, Lee MJ, Williams JL, Nalezny JM, Gee P, Benjamin MB, Farr SB. 1995. The CAT-Tox (L) assay: A sensitive and specific measure of stress-induced transcription in transformed human liver cells. Fundam Appl Toxicol 28:118–128.
Tong AH, Lesage G, Bader GD, Ding H, Xu H, Xin X, Young J, Berriz GF, Brost RL, Chang M, et al. 2004. Global mapping of the yeast genetic interaction network. Science 303:808–813.
USEPA. 2004. Acetochlor: Report of the Cancer Assessment Review Committee (CARC), fourth evaluation. Health Effects Division, Office of Pesticide Programs, Washington, DC. Final report dated 31 August, 2004. TXR No. 0052727.
van der Ven K, Keil D, Moens LN, van Leemut K, van Remortel P, de Coen WM. 2006. Neuropharmaceuticals in the environment: Mianserin-induced neuroendocrine disruption in zebrafish (*Danio rerio*) using cDNA microarrays. Environ Toxicol Chem 25: 2645–2652.
Veldhoen N, Skirrow RC, Osachoff H, Wigmore H, Clapson DJ, Gunderson MP, Van Aggelen G, Helbing CC. 2006. The bactericidal agent triclosan modulates thyroid hormone-associated gene expression and disrupts postembryonic anuran development. Aquat Toxicol 80:217–227.
Vincent R, Goegan P, Johnson G, Brook JR, Kumarathasan P, Bouthillier L, Burnett RT. 1997. Regulation of promoter-CAT stress genes in HepG2 cells by suspensions of particles from ambient air. Fundam Appl Toxicol 39:18–32.
Yauk CL, Berndt ML. 2007. Review of the literature examining the correlation among DNA microarray technologies. Environ Mol Mutagen 48:380–394.
Zhang X, Yu RM, Jones PD, Lam GK, Newsted JL, Gracia T, Hecker M, Hilscherova K, Sanderson T, Wu RS, et al. 2005. Quantitative RT-PCR methods for evaluating toxicant-induced effects on steroidogenesis using the H295R cell line. Environ Sci Technol 39:2777–2785.

3 Application of Genomics to Tiered Testing

*Charles R Tyler, Amy L Filby, Taisen Iguchi,
Vincent J Kramer, DG Joakim Larsson,
Graham van Aggelen, Kees van Leeuwen,
Mark R Viant, and Donald E Tillitt*

CONTENTS

3.1 Background and Regulatory Framework 34
    3.1.1 Tiered Testing 34
    3.1.2 Effects Assessment 34
    3.1.3 Exposure Assessment 35
    3.1.4 Rationale of the Tiered Approach 35
3.2 Intelligent Testing Strategies and Genomics 36
3.3 Genomics and Tiered Testing 38
    3.3.1 Single Biomarkers 38
    3.3.2 Genomics 40
        3.3.2.1 Transcriptomics 40
        3.3.2.2 Proteomics 41
        3.3.2.3 Metabolomics 42
        3.3.2.4 Fingerprinting and Profiling 43
3.4 Potential Applications of Genomics in Tiered Testing 44
3.5 Animal Species 47
3.6 Cost Benefits and Reduction of Animal Usage in Applying Genomics
    in Tiered Testing 47
    3.6.1 Genomic Cost 47
    3.6.2 Vertebrate Animal Reduction 49
3.7 Application of Genomics into Tiered Testing: What, When, and
    Associated Research Needs 50
3.8 Bioinformatics 53
3.9 Regulatory Challenges and Recommendations 54
References 55

3.1 BACKGROUND AND REGULATORY FRAMEWORK

This chapter focuses on the development and application of genomics to tiered testing. It attempts to capture the potential utility of genomics for enhancing tiered testing, including how genomics may be used to support streamlining of the present chemical testing process. The chapter addresses the immediate and longer term goals for genomic tools in tiered testing, including identifying the associated research needs, and addresses the likely impacts (and longer term benefits) on animal testing, and the associated financial costs. Finally, the chapter presents a series of regulatory challenges and recommendations for the development and implementation of genomics into tiered testing. However, to appreciate how genomics might be developed for, and applied to, tiered testing requires an appreciation for how tiered testing works in the risk assessment of chemicals. The initial part of this chapter, therefore, sets out the present (and evolving) framework for this.

3.1.1 TIERED TESTING

Risk assessment of chemicals is a tiered, stepwise process distinguished by levels of increasing complexity, beginning with the preliminary categorization step, followed by a refined or screening assessment, and progressing to the full, comprehensive risk assessment (OECD 1992; CEC 1994; van Leeuwen and Hermens 1995; Van Leeuwen et al. 1996). For each tier, a minimum level of information is required about the substance, including its chemical nature, persistence, and production volumes. For surveying high-production-volume (HPV) chemicals for potential effects, the Organization for Economic Cooperation and Development (OECD) has established an international program on screening information data sets (SIDS) that includes the basic information needed to perform a preliminary assessment of a substance's potential risk (OECD Existing Chemicals Program; www.oecd.org).

The stepwise or tiered process takes into consideration different routes of exposure for all organisms, up to and including top predators at the highest level of the food chain. Risk assessment at each tier seeks to judge the probability of adverse effects in the environment due to exposure levels predicted (or measured) to arise from the use or release of the substance into the environment. The tiered testing schemes, which may differ in shape and form, are applied in the assessment of industrial chemicals, pharmaceuticals, pesticides, and other natural and synthetic substances used in commerce. There are several risk assessment schemes employed by regulatory agencies throughout the world. In general, these schemes compare (no) effect levels to exposure levels and may conclude that: (a) unacceptable risk exists if exposure levels approach or exceed effect levels, or (b) acceptable risk exists if exposure levels are sufficiently below effect levels.

3.1.2 EFFECTS ASSESSMENT

The refinement of effects approach to hazard and risk assessment was first described more than 20 years ago and has been adopted within regulatory schemes around the globe (Kimerle et al. 1977; Cairns et al. 1979; Hansen et al. 1999). The tiered testing approach was envisioned to proceed from screening, to acute and chronic bioassays, to microcosm and mesocosm testing in order to provide more realistic and refined effects assessments by addressing higher levels of biological organization. Tier 1 testing may use simple structure–activity relationships and/or receptor binding assays for the mechanisms of concern. The Tier 1 level of testing is still basic (acute or short-term toxicity) and often derives nothing more than the effective lethality dose or concentration for a chemical. It uses assessment factors to predict and protect for chronic effects. The first tier of effects assessment for aquatic habitats normally covers short-term tests with fish, daphnids, and other aquatic invertebrates and algae; for terrestrial habitats, it utilizes short-term tests in birds and mammals, earthworms, foliage-dwelling arthropods, and terrestrial plants.

In Tier 2, chronic toxicity tests are generally performed and the focus is on sensitive species and life stages—for example, embryo-larval stages of fish—and on critical processes such as growth, development, reproduction, and possibly behavior. Other species may be additionally included, depending on the fate and exposure routes. Tier 3 includes longer term tests such as life cycle and multigeneration studies. At the highest level (Tier 4), population and community level effects are assessed through actual monitoring of wildlife populations and using microcosm and mesocosm experiments.

Regulatory agencies often apply assessment factors, also known as safety factors or extrapolation factors, to the effects endpoints (LD50 [lethal dose 50%], LC50 [lethal concentration 50%], EC50 [effective concentration 50%], NOEC [no observed effect concentration]) in order to protect untested species that may be more sensitive or more desirable (such as rare, threatened and endangered species) than the species tested in the laboratory. The aim is to derive so-called PNECs (predicted no effect concentrations) for ecosystems. This PNEC is compared to the actual or predicted exposure levels in order to assess the risks.

3.1.3 Exposure Assessment

Running in parallel with the tiered testing of ecotoxicological effects are exposure assessments. At the level of Tier 1, production and release volumes and simple fate (very often dilution) processes are considered. At Tier 2, there is a greater realism for the distribution and fate processes of the chemicals. Fate is considered together with available environmental data, including effects of weather and hydrogeology on fate. Very often, multimedia models are used to predict environmental fate and transport. At Tier 3, site-specific models are often used and studies can be developed and applied to calibrate those models. For the highest tier, field monitoring is employed to assess distributions of actual exposure and provide temporal and spatial resolution. Similar to the effects evaluation, at each level of exposure evaluation greater experimental information is needed on the physicochemical and environmental fate characteristics of the substance being evaluated (e.g., hydrolysis, photolysis, adsorption and desorption, solubility, etc.). This includes determination of potential bioavailability, bioaccumulation and/or biomagnification factors, normally at Tiers 3 or 4.

3.1.4 Rationale of the Tiered Approach

The main reason why tiered testing procedures are applied in risk assessment is to improve regulatory efficiency. Because tens of thousands of natural and synthetic chemical substances are currently used in the global economy, regulatory agencies must be efficient in their risk assessment methods to adequately protect the environment and to maintain global economic productivity. Reductions in vertebrate test organism usage may also be realized from tiered approaches to chemical testing.

Nevertheless, in order to perform a reasonably complete risk assessment of a chemical substance, a number of studies have to be performed and subjected to critical analysis. For example, the risk assessments performed for individual substances in the context of the European Union (EU) Existing Chemicals Program typically cover some 300 pages, give as many references to scientific studies, and are the results of several years of committee work. Pesticide assessments can require the use of thousands of vertebrate animals (Mattsson et al. 2003) and the dossiers themselves can reach lengths of thousands of pages. Certainly, regulatory assessments of pesticides must be rigorous, but can they be done more intelligently both in terms of regulatory efficiency and animal testing? The same question holds for the assessment of pharmaceuticals (TSCA 1976; The ChemRTK Initiative 1998; CEPA 1999; EMEA 2005). Additionally, for a significant number of existing chemical substances, there is a paucity of effects and exposure information (NRC 1984; USEPA 1998; Allanou et al. 1999) that could be supplemented using tiered testing approaches.

3.2 INTELLIGENT TESTING STRATEGIES AND GENOMICS

More than 230,000 chemical substances are identified by the American Chemical Society's Chemical Abstracts Service as inventoried or regulated substances (www.cas.org/cgi-bin/regreport.pl). For many, the available information on their associated risk to the environment is limited and, specifically, may be inadequate to address long-term risk. The potential number of substances to be assessed for effects increases dramatically when the products of degradation are also considered. Except for pesticide assessments, degradation products are neglected in most chemical assessment schemes and their toxicity is left unassessed. Balancing this, however, most degradation products are likely to be less biologically active than the parent because many, but not all, degradation processes increase solubility, increase polarity, and decrease lipophilicity.

It is proposed that the most efficient way to carry out hazard and risk assessments of large numbers of chemical substances, while minimizing costs to society and vertebrate animal testing, is to obtain the necessary information by means of so-called intelligent testing strategies (ITS; van Leeuwen et al. 2006). These strategies are integrated approaches comprising multiple elements aimed at increasing the efficiency of the risk assessment process (Bradbury et al. 2004). While the details of the different proposals for ITS vary, a number of common components can be identified:

- screening with in vitro tests
- testing with optimized in vivo assays
- identifying thresholds of toxicological concern
- developing and utilizing structure–activity relationships (SARs) and quantitative structure–activity relationships (QSARs)

Application of Genomics to Tiered Testing

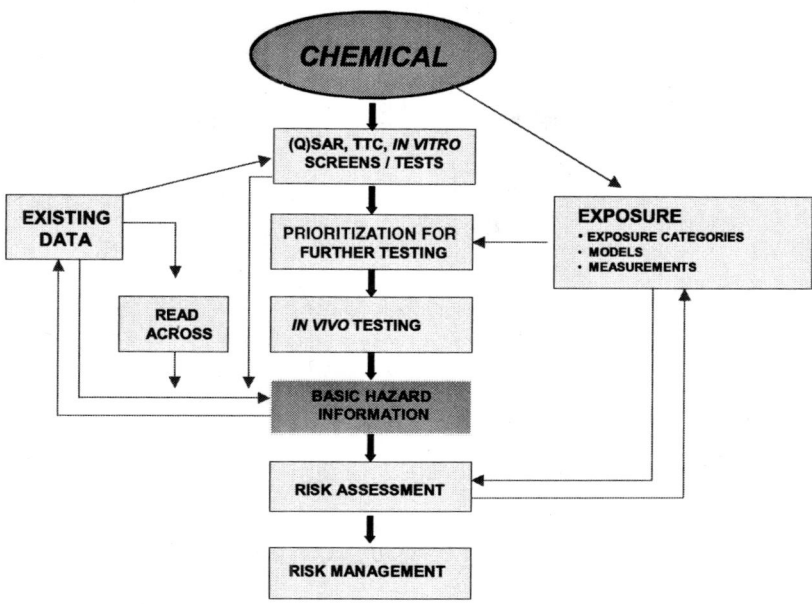

FIGURE 3.1 Framework illustrating how combining use and exposure information and effects information obtained from (quantitative) structure–activity relationships ([Q]SARs), read-across methods, thresholds of toxicological concern (TTCs), and in vitro tests prior to in vivo testing can provide a more rapid, efficient, and cost-effective way to perform risk assessment of chemicals. (Taken from Bradbury SP et al. 2004. Environ Sci Technol 38:463A–470A. Reproduced with permission from Environ Sci Technol. Copyright 2004 Am Chem Soc.)

- clustering and read-across (the transfer of the hazard profile of one chemical to another with a similar structure) of chemical categories
- assessing exposure leading to exposure-based waiving of further testing

None of these components is more important than any other, since the ultimate aim is to obtain reliable information on the properties of chemical substances with minimal use of vertebrate animals. In principle, the information could be obtained in multiple ways by means of different combinations of the components. However, some ways could be more efficient than others, depending on the underlying rationale of the strategy (e.g., rapidly evaluating alternative pesticides). An example of a generic testing strategy is depicted in Figure 3.1 (taken from Bradbury et al. 2004). Similar schemes have been published by other authors (e.g., Health Council of the Netherlands 2001; Combes et al. 2003; Hofer et al. 2004). Some strategies have been proposed for the prioritization of substances for further assessment, while others are endpoint-specific strategies designed for the assessment of particular endpoints (e.g., Worth et al. 1998; Worth 2004).

A major challenge in the next decade is to achieve regulatory acceptance of (Q)SARS, in vitro tests, read-across, and, of most immediate relevance to this chapter, application of the genomic technologies to help refine and optimize ITS approaches (Bradbury et al. 2004; Fentem et al. 2004). For regulatory purposes, the elements of ITS (Figure 3.1) and new genomic technologies could potentially be used either to supplement experimental data or to reduce or even replace testing. The use of genomics to supplement experimental data would include contributing to priority setting of chemicals, guiding experimental design (e.g., selection of tests and doses), and providing mechanistic information.

In considering the development and application of ITS for risk assessments of chemical substances, there is a need for rethinking and reframing the current "3-R" strategy of replacement, reduction, and refinement (Russell and Burch 1959). A "7-R" strategy has recently been proposed (van Leeuwen et al. 2006) that includes exposure. In this strategy the focus is on 1) *r*isks, assessed in a 2) *r*epetitive manner (a tiered approach, going from simple to refined or comprehensive risk assessment, if necessary, as advocated by the OECD). The strategy identifies the need to move away from the chemical-by-chemical approach as much as possible and to focus on 3) *r*elatives—that is, on families or categories of chemicals (a groupwise approach), applying (Q)SARs, read-across, and exposure categories. As such, the guiding philosophy is 4) *r*estriction of testing (waiving of testing) where possible and carrying out in vivo testing where needed in order to prevent damage to human health and/or the environment. The overall strategy also encompasses the current 3-R strategy of 5) *r*eplacement, 6) *r*efinement, and 7) *r*eduction of mammalian and avian animal tests. The suggestion here is that a broader 7-R strategy should be developed for REACH (Registration, Evaluation, Authorization, and Restriction of Chemicals; http://ec.europa.eu/environment/chemicals/reach/reach_intro.htm) and other regulatory frameworks by applying ITS, in which the further development and application of genomic technologies be explored as far as possible.

3.3 GENOMICS AND TIERED TESTING

3.3.1 SINGLE BIOMARKERS

Biological responses to toxicants involve changes in normal patterns of gene expression, and the use of transcription of genes as "biomarkers" of chemical exposure has often met with good success (Livingstone 1993; Wong et al. 2000; McClain et al. 2003; Roberts et al. 2005). An example showing that molecular responses can signal effectively for chemical exposure is for hepatic vitellogenin (VTG) messenger RNA (mRNA) induction in fish (Thomas-Jones, Garcia-Reyero et al. 2004) and in some other oviparous vertebrates (Lorenzen et al. 2003; Custodia-Lora et al. 2004; Gye and Kim 2005; Huang et al. 2005). Studies in a wide range of fish species have shown that induction of the VTG gene is highly sensitive to steroidal estrogen. Importantly, in the context of developing molecular approaches to tiered testing, mRNA responses can be detected after relatively short exposure periods (24 to 48 hours; Bowman et al. 2000; Hemmer et al. 2002; Schmid et al. 2002; Thomas-Jones, Thorpe, et al. 2003).

Other single gene biomarkers that effectively signal for exposure to specific classes of chemicals include the vitelline envelope proteins (VEPs, also referred to as zona pellucida or zona radiata proteins), similarly up-regulated by environmental estrogens (Celius et al. 2000; Lee et al. 2002; Knoebl et al. 2004); cytochrome P450 1A (CYP1A), up-regulated by planar aromatic compounds that bind to the aryl hydrocarbon (Ah) receptor (Hahn and Stegeman 1994; Rees et al. 2003); and metallothionein, induced on exposure to and/or accumulation of certain heavy metals (George et al. 1992; Kille et al. 1992; Lam et al. 1998; Tom et al. 2004).

Often the functional significance of these changes in gene expression to the "health" of the organism has not been well established. This is due principally to the lack of extensive and thorough experimentation leading to a detailed understanding of the role of the mRNA signal as it relates to apical (i.e., higher levels of biological organization) adverse outcomes. A recent example of the level of experimentation needed to link mRNA changes to adverse outcomes is reported in a study of zebrafish (*Danio rerio*) demonstrating that suppression of the expression of mRNA transcripts for DMRT-1 and anti-Mullerian hormone as a consequence of exposure to the synthetic estrogen ethinylestradiol (at environmentally relevant concentrations) during early life signaled for a subsequent retardation in testicular development (Schulz et al. 2007).

Proteins are arguably as important (or more important than) as biomarkers of biological function than RNA transcripts because not all mRNA sequences are transcribed and many proteins undergo post-translational modification before becoming physiologically active. A wide range of protein markers have been used to study disease and assess alterations in animal physiology as a consequence of exposure to environmental stressors. As for their transcripts described before, VTG and VEPs have been used as biomarkers of exposure to environmental estrogens (Larsson et al. 1994; Sumpter and Jobling 1995; Celius and Walther 1998), metallothionein for exposure to certain metals (Burkhardt-Holm et al. 1999), and CYP1A protein and enzyme (ethoxyresorufin O-deethylase; EROD) activity to indicate exposure to Ah receptor ligands (Goksoyr and Forlin 1992). For VTG, linkages have been drawn between high circulating concentrations and adverse effects on fish health, including on energetics during early life and on renal function (Herman and Kincaid 1988; Schwaiger et al. 2000; Zaroogian et al. 2001; Liney et al. 2006). High plasma VTG concentrations have also been associated with reduced reproductive success in fish, a population-level relevant endpoint (Kramer et al. 1998; Thorpe and Tyler 2006; Miller et al. 2007)).

Proteins belonging to the heat shock protein (HSP) family have been useful as general indicators of stress (Bradley et al. 1998). However, their low specificity makes them less useful for indicating the source of the stress. Identifying protein markers for human disease is an area of large interest and potential. As an example, by combining information on the levels of three proteins in the cerebrospinal fluid (T-tau, Aβ42, and P-tau$_{181}$), it is possible to predict with good sensitivity and accuracy if a patient will develop Alzheimer's disease within the next few years (Hansson et al. 2006).

Metabolites are the endogenous precursors and products of metabolism and include amino acids, carbohydrates, lipids, and many other metabolite classes. A major benefit of characterizing metabolite concentrations is that the measurements represent a high degree of functionality closely linked to organism physiology. Attempts have been made to align alterations in single metabolites with alterations in normal physiology and adverse toxic effects. Examples include dihydroxynonene mercapturic acid, a urinary metabolite of 4-hydroxynonenal, as a biomarker of lipid peroxidation (Peiro et al. 2005); reduced glutathione in the liver of chub (*Leuciscus cephalus*) as an indicator of oxidative stress (Winter et al. 2005); and testosterone levels as a biomarker of endocrine disruption in the estuarine mysid *Neomysis integer* (Verslycke et al. 2004). In addition, phosphagens have been suggested as indicators of altered energetic status following toxicant exposure—for example, phosphoarginine levels in copper-exposed red abalone (*Haliotis rufescens*) (Viant et al. 2002) and the concentration of phosphocreatine in medaka (*Oryzias latipes*) embryos after exposure to the pesticide dinoseb (Viant, Pincetich, Hinton, et al. 2006). Likewise, ratios of metabolites have been used in a similar manner—for example, the adenylate energy charge that uses the levels of ATP, ADP, and AMP (Thebault et al. 2000).

There are few instances where single genes, proteins, and/or metabolites and their functional significance have been developed to a point where they could potentially be imported into tiered testing in the short term. One exception, however, is induction of VTG, which could now be applied into the lower testing tiers and used as a trigger to initiate higher tier testing for estrogenic disruption. Indeed, the use of VTG induction in this regard is already under consideration for such a purpose through international testing efforts coordinated by the OECD.

3.3.2 Genomics

Some single genes, proteins, and metabolites can be highly effective in signaling for the mode of action (MOA) of a chemical. VTG is a good example as it is induced only under stimulation of the estrogen biosynthesis pathway and/or activation of the estrogen receptor. Most single biomarkers, however, are more limited in their ability to signal for the effect pathways of chemicals. Some of the cytochrome P450 enzymes illustrate this, as they are up- and down-regulated by a wide range of different chemicals that are structurally and functionally diverse (e.g., beta-naphthoflavone: Chung-Davidson et al. 2004; imidazole derivatives: Babin et al. 2005; PCBs: Fisher et al. 2006; steroidal estrogens: Vaccaro et al. 2005; 4-nonylphenol: Vaccaro et al. 2005; antiparasitic drugs: Bapiro et al. 2002).

The functional interpretation of single gene, protein, and metabolite biomarkers of exposure and effect for chemicals relies heavily upon pre-existing knowledge of the pathway of effect. Adding to the difficulty of interpreting functional significance of these single site biomarkers, toxic effects can be mediated by more than one mechanism of action. Therefore, the use of single gene, protein, and metabolite biomarkers should be restricted to hazard identification, rather than risk assessment, of chemicals. Single site biomarkers are presently used generally as mechanistic signposts for directing further testing and not for establishing no observed adverse effect levels (NOAELs) of chemicals for use in risk assessment (Hutchinson et al. 2006). Genomic (transcriptomics, proteomics, metabolomics) technologies offer the potential for a more global approach with considerably greater predictive power for adverse effects than for single biomarker responses.

3.3.2.1 Transcriptomics

Transcriptomics is the large-scale study of the transcriptome—the set of all mRNA molecules (transcripts) of an entire organism, tissue, or cell type. Measurement of the expression profiles of multiple genes is advancing rapidly through the development and application of gene arrays and reverse transcription polymerase chain reaction (RT-PCR) techniques. Quantitative PCR (Q-PCR) has the advantage over arrays in that this, as the name implies, is a quantitative technique. Compared with arrays, Q-PCR is more restricted in the number of genes that can be screened at any one time. Microarray research has been most prevalent for mammalian models for screening and diagnosis of cancers (Wadlow and Ramaswamy 2005) and other diseases or disorders (e.g., Archacki and Wang 2004) in humans. Mammalian arrays have been applied successfully to toxicity testing of pharmaceuticals and chemicals (reviewed in Vrana et al. 2003; Lettieri 2006). Transcriptomics in mammalian models has further been applied to investigate the mechanistic pathways for chemical effects. The platforms most commonly used for this research in mammals are oligonucleotide arrays.

Less research has been forthcoming for microarrays in nonmammalian species. Recently, however, there has been an increase in activity in this area, and the development of large gene sets for a wide range of species has resulted in a suite of research papers that have analyzed transcriptome profiles relative to the physiological state in birds, fish, frogs, and invertebrates (van Hemert et al. 2004; Baldessari et al. 2005; Kucharski and Maleszka 2005; Soetaert et al. 2006). Examples of this in fish include the effects of temperature in carp (*Cyprinus carpio*; Gracey et al. 2004) and rainbow trout (*Oncorhynchus mykiss*; Vornanen et al. 2005), the effects of stress in rainbow trout (Krasnov et al. 2005), and transcriptomic profiles during development and between sexes in zebrafish (Mathavan et al. 2005; Santos et al. 2007). Genome projects for zebrafish (The Sanger Institute; www.sanger.ac.uk/Projects/D_rerio), medaka (National Institute of Genetics; http://dolphin.lab.nig.ac.jp/medaka/), *Daphnia* (DOE Joint Genome Institute; http://cgb.indiana.edu/genomics/projects/7), *Xenopus laevis* (DOE Joint Genome Institute; http://www.jgi.doe.gov/xenopus), and *Xenopus tropicalis* (DOE Joint Genome Institute; http://www.jgi.doe.gov/xenopus) and the availability of very extensive sequence information for carp (www.ncbi.nlm.nih.gov.uk) have greatly facilitated transcriptomic research for these species used in regulatory toxicology. Other species relevant to regulatory toxicology for which there has been a rapid development of genome information include the fathead minnow (*Pimephales promelas*; NCBI Genbank www.ncbi.nlm.nih.gov.uk), and stickleback (*Gasterosteus aculeatus*; http://cegs.stanford.edu/index.jsp; Genome Sciences Centre; http://www.bcgsc.ca/lab/mapping/stickleback).

3.3.2.2 Proteomics

Proteomics is the large-scale study of the proteome—the set of all proteins of an entire organism, tissue, or cell type. There has been a rapid evolution of proteomic methods capable of providing broad characterizations of proteins expressed within cells, organs, or, in some instances, whole organisms, including species relevant to ecotoxicology research (e.g., zebrafish, Schrader et al. 2003). There have been large expectations placed on proteomic approaches to find novel predictive biomarkers for human disease as well as biomarkers for use in ecotoxicology. Proteomics has been used successfully to develop human medical diagnosis. For example, protein markers that predict neurodegenerative diseases have been identified in cerebrospinal fluid (Ruetschi et al. 2005). Recently, proteomics has also been used to classify histologically defined hepatocellular adenomas in a marine flatfish, the dab (*Limanda limanda*; Stentiford et al. 2005).

It has yet to be demonstrated, however, to what extent proteomic approaches can be used for providing diagnostic power in an ecotoxicological context. This limited success in identifying novel biomarkers in ecotoxicology research is, to some extent, linked with shortcoming in the proteomic analytical approaches in predominant use today. Methods vary, and they typically include protein isolation and separation steps with techniques like 2-D gel electrophoresis or high-pressure liquid chromatography, followed by mass spectrometry (MS) analyses to identify peptide profiles or amino acid composition as a basis for identification of specific proteins (Aebersold and Mann 2003). Two-dimensional gel electrophoresis linked to image analyses and the subsequent identification of the isolated proteins using MS is still a relatively tedious approach and limited in the number of proteins that can be practically studied.

The scarcity of sequence information for many test and sentinel species has also hampered the development of proteomics in ecotoxicology. Surface-enhanced laser desorption and ionization time of flight mass spectrometry (SELDI-TOF-MS), protein-antibody arrays, isotope-coded affinity tagging (ICAT), and high-resolution Fourier transform ion cyclotron resonance (FT-ICR) MS are techniques that provide higher coverage of proteins and have better promise for standardization for testing applied to regulatory purposes. Probably the most important use of proteomics in ecotoxicology in the immediate future will be to provide linkages from gene to phenotype that can improve our understanding of the modes and mechanisms of action by which different chemicals or mixtures act (Albertsson et al. 2007).

3.3.2.3 Metabolomics

Metabolomics is the large-scale study of the metabolome—the totality of metabolites (low molecular weight metabolites involved in all biological reactions required for normal function of a tissue or whole organism) in an organism, tissue, or cell type (Nicholson et al. 2002; Schmidt 2004; Robertson 2005). The metabolome includes a wide range of compounds ranging from polar organics (e.g., amino acids, small peptides, glucosides) to comparatively nonpolar lipids to inorganic elements. It has been shown that endogenous metabolite profiles can be used as a diagnostic tool of physiological status (Nicholson et al. 2002) and thus metabolomics has the capability to capture a more integrated assessment of the organism's well-being than transcriptomics or proteomics (Robertson 2005).

Different high-resolution MS and nuclear magnetic resonance (NMR) spectroscopy techniques provide the primary basis for generating metabolomic data (Dunn and Ellis 2005). A number of reports have related suites of metabolites to whole organism health, but most of these have focused on aspects of human health conducted in humans, rats, and mice (Brindle et al. 2002; Burns et al. 2004; Yang et al. 2004; Odunsi et al. 2005). A significant body of this research has been conducted by Nicholson and colleagues (Imperial College, London, United Kingdom), the pioneers of the NMR-based approach (Lindon et al. 2000; Nicholson et al. 2002). In a collaboration with 6 pharmaceutical companies, Nicholson and colleagues generated comprehensive metabolic data using ^{1}H NMR spectroscopy of urine and serum from rats exposed to model toxic chemicals. In total, approximately 80 model toxic chemicals were examined, resulting in the generation of the first highly developed systems for the prediction of liver and kidney toxicity. The successes achieved in this project The Consortium for Metabonomic Toxicology (COMET; Lindon et al. 2003, 2005) lends significant weight to the possibility of implementation of metabolomics technologies into (tiered) risk assessment.

Metabolomics has recently been applied to assessments of organism health in nonmammalian animals, including aquatic species relevant to ecological risk assessments (e.g., Viant, Werner, et al. 2003). Examples of this include the classification of withering syndrome in the red abalone, using metabolic fingerprints derived from ^{1}H NMR spectroscopy (Viant, Rosenblum, et al. 2003; Rosenblum et al. 2005), and the application of MS-based metabolomics to classify histologically defined hepatocellular adenomas in dab in which specific markers associated with carcinogenesis were identified (Stentiford et al. 2005). Other metabolomics studies have been used to identify biomarker patterns associated with specific chemical insults, such as in earthworms (*Eisenia veneta*) exposed to copper and fluorinated anilines (Bundy et al. 2002, 2004), medaka embryos exposed to trichloroethylene and dinoseb (Viant et al. 2005; Viant, Pincetich, Hinton, et al. 2006), in chum salmon (*Oncorhynchus tshawytscha*) embryos exposed to pesticides (Viant, Pincetich, and Tjeerdema 2006), and in rainbow trout exposed to ethinylestradiol (Samuelsson et al. 2006).

3.3.2.4 Fingerprinting and Profiling

Two strategies have evolved in genomics research, fingerprinting and profiling, which can be differentiated according to whether the endpoints measured (transcripts, proteins, metabolites) are identified or not. Here, we define genomic-based fingerprinting as an unbiased approach to measure as many endpoints as possible, where these endpoints are not identified (or do not need to be identified). This term is often used synonymously with signature and pattern, but here we propose to formalize the term fingerprinting to apply across all the genomics technologies (see Section 3.9, "Regulatory Challenges and Recommendations"). The obvious advantage of fingerprinting is that, because sequence annotation or metabolite identification is not required, a wide range of organisms can be studied.

For the case of transcriptomics, this opens up the possibility of including a wider range of species because extensive sets of ESTs (expressed sequence tags) are more readily available than partial or fully annotated genomes for organisms of ecological relevance. The rationale behind all genomic fingerprinting approaches is that specific mechanisms of action will result in unique patterns in the fingerprint that are significantly different from the fingerprint of a "healthy" control organism. Fingerprinting can, therefore, be used to rapidly define dose–response relationships, where increasing doses of chemicals will induce increasing deviations from the normal phenotype. Using this approach within a multivariate framework of analysis (Viant et al. 2005), a no-observable-effect level (NOEL) based upon multiple endpoints can be derived.

Genomic-based profiling is an approach in which the measured endpoints are unambiguously identified. In the case of transcriptomics, this requires that sequence annotation of the surrogate species is known. Although this is true for several species used in risk assessment, there are a number of key organisms (e.g., fathead minnow) for which the full genome sequence is urgently required; only when this is achieved can the full potential of transcriptomic profiling be realized. For the case of metabolomics, profiling requires the construction of a library of metabolite standards for peak annotation. Although this is a major undertaking (the cellular metabolome is estimated to include many thousands of metabolites, of which a large proportion may be unknown), this resource would not be species specific.

3.4 POTENTIAL APPLICATIONS OF GENOMICS IN TIERED TESTING

Transcriptomics, proteomics, and metabolomics measure responses at different levels of biological organization, and each provides a different insight into the biochemical or molecular status of an organism. All three approaches have potential for defining toxicity pathways and use in tiered testing, and while each of the techniques has its own challenges, none can yet be considered routine in the context of applying it to ecotoxicology and risk assessment. The authors of this chapter feel that transcriptomics (microarrays) and metabolomics (NMR or MS) are toxicogenomic techniques that currently show the most immediate promise for application to tiered testing.

Changes in the transcriptome (changes in gene expression) are often rapid and signal for an altered phenotype, while metabolomics provides an integrative signal of chemical effect and subsequent processing. Used in tandem, these two techniques have the capacity to enhance very significantly our understanding of chemical effect measures. At this time, proteomic techniques are not well suited for the global fingerprinting approaches that have the most promise for routine testing in the near term. Proteomics could add mechanistic information; however, in development of a tiered testing scenario, overall it is felt that resources may be better directed into transcriptomics and metabolomics to advance the mechanistic understanding of ecotoxicology and to streamline workload and resource utilization in government-mandated testing programs.

Genomics might be usefully applied in hazard identification for chemicals, as depicted in a generic tiered testing paradigm (Figure 3.2). At the screening stage, genomics could be used to support priority setting of chemicals. Used in combination with in vitro cultures of available cell lines, with primary cell cultures (e.g., hepatocytes), and possibly nonmammalian cell lines, chemical fingerprinting could serve to identify chemicals eliciting responses known to be correlated to adverse outcomes and thus prioritize them for higher tier testing. Screening at this level with genomic tools is likely to be especially applicable for chemicals that have well-defined MOAs (such as pharmaceuticals). Additionally, approaches at this level are amenable to high-throughput analyses at reduced costs.

At Tier 1, in acute toxicity tests, the genomics technologies could generate information on possible MOAs of the test chemicals and thus better direct the selection of tests and chemical dosing in higher tiers of testing. Better definition of MOA would not only help focus test resources, but also could provide information critical

Application of Genomics to Tiered Testing

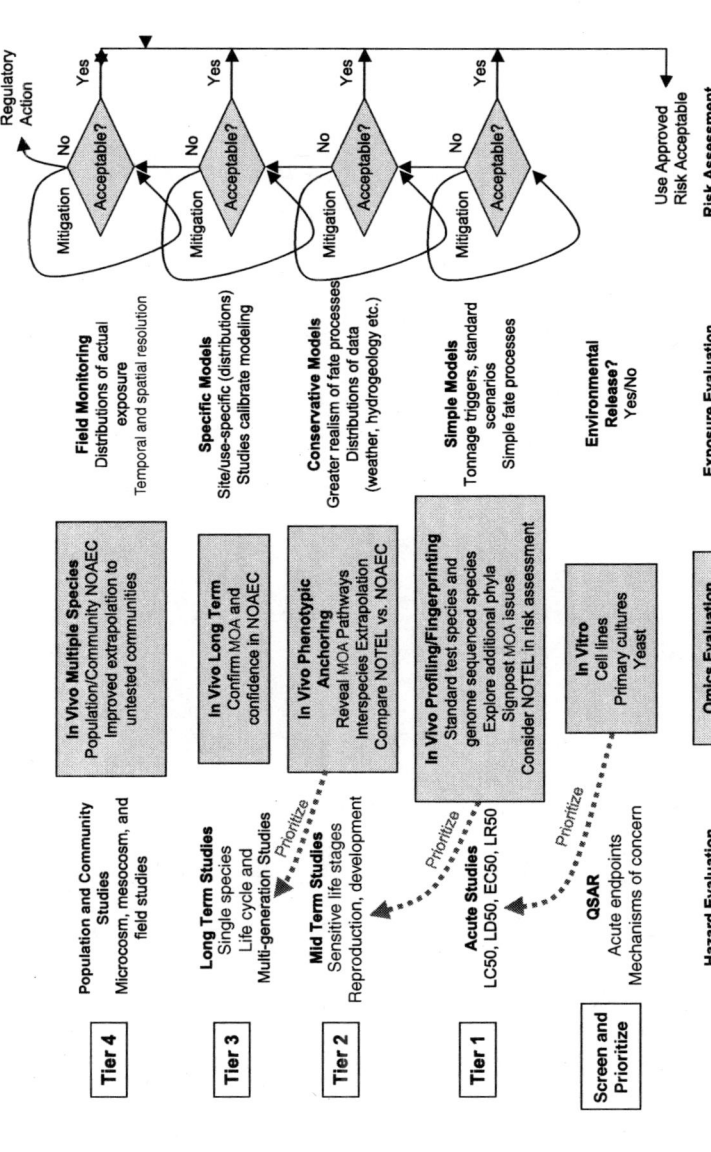

FIGURE 3.2 Generic tiered-testing and risk assessment framework for prospective ecological risk assessments indicating where toxicogenomic data could be incorporated relative to other models. Rapid prioritization arrows indicate where genomics could be used to inform on hazard assessment and better direct higher tiers of testing. Abbreviations: MOA, mode of action; NOTEL, no observable transcriptional effect level; NOEC, no observable effect level; NOAEC, no observable adverse effect level. (Taken from Bradbury SP et al. 2004. Environ Sci Technol 38:463A–470A. Reproduced with permission from Environ Sci Technol. Copyright 2004 Am Chem Soc.)

to reducing uncertainty in risk assessments. For example, a significant source of concern for risk assessments involves prediction of the toxicity of chemical mixtures. Genomic profiles could be used to identify compounds with similar versus dissimilar MOA and/or to inform on chemical selection for studies in mixture toxicity models. Profiling of molecular responses in vivo at Tier 1 would build up a picture of mechanistic pathways for the various groups of chemical substances. Based on comparisons of the mechanistic pathways of chemicals in different species or phyla, the use of toxicogenomic techniques could also help establish those situations where extrapolation of chemical effects from one species (or other phylogenetic group) to another is appropriate and warranted.

Phenotypic anchoring will be a key element for the use of genomics in testing at Tier 2. The linkages developed between the molecular responses of the genome and effects observed in the phenotype are extremely useful for determination of the MOA of a chemical (Vrana et al. 2003). The phenotypic response in toxicological terms is the expression of adverse and nonadverse effects in the organism. Chemical substances have the potential to exert toxicity via multiple pathways. At present, the approach used to try to ensure that unanticipated toxicity pathways are realized is to use a number of different types of tests with multiple species and endpoints. This is, of course, expensive and time consuming and can result in large amounts of data that are redundant for the subsequent decision-making process. Insights gained through genomic analysis could serve as a basis for customizing test designs and endpoints such that suites of assays are optimized for a given class of chemicals (the application of ITS). The use of genomics to better target MOA in this type of situation would help focus the investment of resources into tests that would impact risk assessments, rather than generate costly data of little utility. These more detailed analyses would further support information for species extrapolation.

Progressing to Tiers 3 and 4, the use of genomics could better confirm the MOA of the test chemical (and a more certain analysis of the NOAEL) and provide greater certainty for extrapolation from the genomics at lower tiers of testing.

With the appropriate transcriptomic data in conjunction with existing tests to ascertain whether or not a chemical has the potential to elicit toxicity via a particular MOA, a no-observed-transcriptional-effect level (i.e., NOTEL; for mRNA responses) might be developed and applied to tiered testing. Similar statistical concepts could be developed for metabolomic and proteomic responses. To elaborate further, if there is no evidence of interaction of the test chemical with the pathway of concern in terms of gene (or metabolite) expression, this could serve as the basis of a decision for no further testing. A number of statistical analysis issues would have to be addressed as this concept is developed, including the control of the false discovery rate (Reiner et al. 2003) when comparing thousands of gene transcript, metabolite, or protein responses to control levels and accounting for the potential intercorrelation of these responses due to the underlying physiological networks that functionally link them together (Thomas and Ganjii 2006).

At this time, this concept would only be advocated for chemicals with a known or suspected MOA and thus to screen or test for defined pathways of effect, rather than in a broader fashion. It should also be emphasized that this concept could be applied only if the responses to the chemical at the molecular level are equally as sensitive as, or more sensitive, than apical responses observed in the test organism, which may or may not be the case. It is important, however, that the nature of the substance under test is taken into consideration in these assessments. For example, some chemicals may not accumulate to a sufficient degree in short-term tests to elicit a molecular response, but in longer term tests may accumulate and exert an adverse biological effect.

On the other hand, short-term genomic assays may be able to detect and respond to the molecular properties of those substances with a propensity to bioaccumulate. At the highest testing tier in the monitoring of wildlife populations, the absence of a precharacterized molecular profile for a particular chemical substance (or class of chemical substances) in the species of interest would provide strong evidence that the substances are either not bioavailable or below thresholds of transcriptional or metabolomic effects, even though they might be detectable in the environment via analytical chemistry.

3.5 ANIMAL SPECIES

In the context of higher tiered testing, because transcriptomics and proteomics (but not metabolomics) require a bank of sequence information, they are more likely to be applied in the immediate term to the surrogate or model organisms and/or other organisms for which there is substantial sequence information. As the application of genomics is biased to organisms with a sequenced transcriptome or genome, this provides a major incentive for the sequencing of all key species commonly used in tiered testing. Commercial arrays are already available for some of these key model test organisms used in OECD, the Environmental Protection Agency (EPA), and Japan, including *Xenopus laevis*, medaka, zebrafish, fathead minnow, mouse, and rat. Furthermore, there are privately held arrays for a variety of other organisms relevant to tiered testing, including for rainbow trout, salmon, carp, and *Daphnia*.

A fully sequenced genome, however, should not necessarily be a prerequisite for the use of genomics in tiered testing as even a few hundred gene sequences that represent key functional pathways could be most valuable in guiding effect pathway determination. Given the rapid advances in sequence databases, knowledge on primer design, and high-throughput techniques in molecular cloning and sequencing, it is within the capability of most established molecular laboratories to obtain large suites of target genes in the species of interest in a relatively short period of time. Thus, the required resources for genomic studies in sentinel (wildlife) species used in environmental monitoring can be reasonably developed in the short term and may be extended to phyla not commonly included in regulatory testing schemes.

3.6 COST BENEFITS AND REDUCTION OF ANIMAL USAGE IN APPLYING GENOMICS IN TIERED TESTING

3.6.1 Genomic Cost

Traditional toxicological evaluations used in the tiered testing framework for chemicals for human health and environmental endpoints are a major expense to registrants.

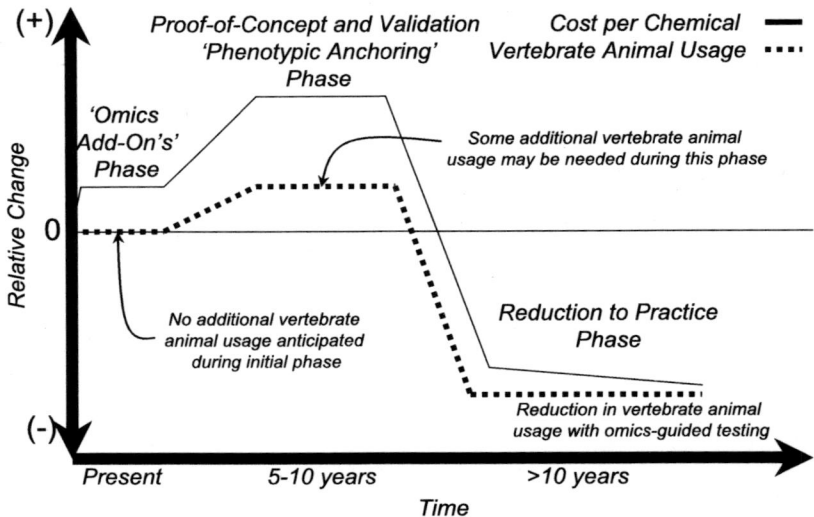

FIGURE 3.3 Conceptual timeline and relative resource requirements for the development and integration of toxicogenomic data into regulatory programs. (Taken from Bradbury SP et al. 2004. Environ Sci Technol 38:463A–470A. Reproduced with permission from Environ Sci Technol. Copyright 2004 Am Chem Soc.)

The European Commission (EC) September 2003 REACH program estimates the direct cost of this testing to be €2.4 billion (2003) over an 11-year period. Based on the approaches and premises outlined earlier in this chapter, it is anticipated that the unit cost of conducting tiered testing by the implementation of genomics technologies and their associated endpoint measurements will, in the long term, be a more cost-effective alternative; of course, it is difficult at this time to predict the actual cost savings.

A review of the EC's 2004 report, "Alternative Approaches Can Reduce the Use of Test Animals," concluded that significant savings can be achieved by applying alternative approaches to traditional testing regimes. In fact, the anticipated savings by applying ITS provides an interesting scenario through which to get prospective cost savings (through reduced vertebrate animal testing) that could be realized from the implementation of genomic technologies. It is estimated that a saving of €1.5 billion would be achieved just by implementing ITS. We propose that genomic technologies have the potential to reduce this cost even further.

A comprehensive genomics database must first be established, however, for these cost savings to become realized. The establishment of this database is critical if genomic tools are to gain full implementation and acceptance into the tiered testing arena. It is envisioned that this database would be developed in concert, over a 3- to 5-year period, with the traditional toxicological approaches to tiered risk assessment (Figure 3.3). Initial incorporation of genomic data into existing regulatory frameworks is expected to cause some increase in resource use, primarily for generating fingerprinting and profiling data using samples derived from existing test schemes. In the slightly longer term (5 to 10 years), resources used (including animals) could increase further as issues associated with phenotypic anchoring and validation are addressed. From this data bank, linkages between genomics endpoint measurements from both transcriptional or metabolomic testing would be made with and validated against those measurable deleterious effects in the targeted organism. That is, the toxicological endpoints routinely measured in tiered testing would be evaluated with respect to the genomic fingerprint and profiles.

Achievement of this phenotypic anchoring of genomic responses certainly requires a commitment of significant fiscal resources, and a logistical coordination among laboratories would be imperative. However, cost savings could be realized in the tiered testing program once the relationship and predictive power of the genomics endpoints are established. Regulatory agencies will need to agree to accept genomics data in lieu of traditional testing schemes for the efficiency of genomics to be fully realized. A potential alternative scenario, however, could be that genomic information is only considered by regulatory agencies to be supplemental to the traditional testing schemes. Consequently, in the end, even more testing is ultimately required. It is hoped that this potential alternative scenario would not come to fruition. A conceptual illustration of the anticipated time line for the realization of fiscal return is provided in Figure 3.3.

3.6.2 Vertebrate Animal Reduction

It is widely advocated in North America, and especially in Europe, that there is a need to reduce (or to eliminate) vertebrate testing, and the application of genomics-based tools has significant potential to help to achieve this goal. It is difficult to predict the actual reduction in numbers of vertebrate animals used in tiered testing, but perhaps the best example of how genomic tools could further reduce vertebrate animal testing is provided by reviewing the EC calculations from their REACH program projections of animal use from implementing ITS. The calculation of the number of vertebrate test animals needed as a consequence of REACH has been carried out by using the approach employed by Pedersen et al. (2003). In this approach, the total amount of data required under REACH was established based on the information requirements of the legislative proposal. Against this, the amount of currently available data, as well as the data already promised under various programs, was determined.

The impact on the testing needs of ITS was assessed assuming three scenarios: 1) a standard scenario representing an average situation regarding acceptance of these methods, 2) a scenario based on minimum acceptance, and 3) a scenario with maximum acceptance. The currently available and expected data, as well as the effect of the use of these methodologies, were then considered in conjunction with the total quantity of data required under REACH, resulting in the estimated test requirements. The number of vertebrate test animals needed for the individual endpoints was established by consulting test laboratories and this was multiplied by the estimated number of studies required under REACH, resulting in estimates of the total amount of test animals that will potentially be needed for the implementation of the REACH legislation (Van der Jagt et al. 2004).

The results of these calculations show that approximately 3.9 million additional vertebrate test animals could potentially be used as a consequence of the introduction of REACH if the use of alternative methods is not accepted by regulatory authorities. However, a considerable reduction in vertebrate animal use can be obtained if genomic techniques were applied more intensively. The standard scenario based on average acceptance of these methods indicates potential savings of 1.3 million test animals. Maximum acceptance of these techniques would even enhance this saving potential to 1.9 million test animals. These savings can be obtained by introducing and accepting methods that are, to a large extent, available today.

In the short and midterm, development and validation of genomics into tiered testing would incur additional vertebrate animal usage. In the short term, the requirements of vertebrate animals and tissues for fingerprinting and profiling could largely be derived from existing test schemes, so the increase in sample requirements is likely to be minimal. In the midterm, however, vertebrate animal use probably would increase, notably to allow the development of phenotypic anchoring. A conceptual illustration of the anticipated impacts on vertebrate animal usage of the development and application of genomics in chemical testing with the potential associated time line is given in Figure 3.3.

3.7 APPLICATION OF GENOMICS INTO TIERED TESTING: WHAT, WHEN, AND ASSOCIATED RESEARCH NEEDS

The application of genomics technologies to regulatory strategies for tiered testing of chemicals requires additional research in both a short- and long-term context. Targeted research efforts are needed for successful application and integration of transcriptomics, proteomics, and metabolomics into tiered testing protocols. The utility of these genomic technologies in the characterization and prioritization of chemicals for hazard testing is only going to be fully recognized after some of these specific research needs are met. Some elements of the genomics technologies can probably be applied to regulatory strategies for tiered testing of chemicals in the short term (within the next 2 to 3 years). Others, however, require considerable development and are medium (3 to 5 years) to longer term prospects (5 to 10 years). This section of the chapter considers genomics that might be developed and applied into tiered testing in these time intervals (Figure 3.3) and the associated research activities needed to achieve these goals.

Mammalian species used in tiered testing protocols for environmental risks of chemicals have the distinct advantage of also being used in testing and evaluation of chemicals for human risk assessment; consequently, the genomics tools for mammalian species are more fully developed than species used in ecological risk assessments. Furthermore, array platforms have a greater standardization. Substantial sequence information is available for some species used in ecological risk assessment and, as detailed earlier, commercial oligonucleotide arrays are already available for some of these (e.g., zebrafish). For the most part, however, the array platforms applied for species used in ecological risk assessment vary (and include both cDNA and oligonucleotide arrays). Furthermore, even for species where the genome has been sequenced, the fidelity of the annotations of these genomes is not always robust (e.g., zebrafish). Therefore, an immediate need for application of genomics technologies to tiered testing is that of standardized tools for the relevant nonmammalian species. With concerted international support, it would be possible for this requirement to be achieved rapidly.

Fingerprinting approaches for transcriptomics using EST-based arrays and, for metabolomics, NMR spectroscopy of MS offer methods for application into tiered testing in the very near future (next 2 to 3 years); these could add significant benefit in terms of streamlining the whole testing process as shown in Figure 3.2. Fingerprinting is the most widely used approach in metabolomics at present, predominantly because libraries of metabolite standards required for peak annotation and subsequent profiling techniques do not yet exist in the public domain (see later discussion). Both ^{1}H NMR spectroscopy and MS have been used to fingerprint the metabolic phenotypes of multiple species across multiple phyla (e.g., terrestrial [Gibb et al. 1997] and aquatic [Viant, Rosenblum, and Tjeerdema 2003] invertebrates, fish [Viant, Werner, et al. 2003; Solanky et al. 2005; Viant et al. 2005; Viant, Pincetich, Hinton, and Tjeerdema 2006; Viant, Pincetich, Tjeerdema 2006], and wild mammals [Griffin et al. 2000, 2001]). The NMR approach benefits from a rapid, reproducible, and unbiased analysis of metabolites, but is limited to the detection of only relatively abundant metabolites, including the amino acids, some carbohydrates, many organic acids and bases, and relatively nonspecific signals from lipids. This platform is stable, however, and has been shown to yield a high degree of reproducibility between samples analyzed in different laboratories in the United Kingdom, Europe, and the United States with spectrophotometers of different capabilities. This is of critical importance to reliable analysis.

As applied to metabolomics, MS has a considerably higher sensitivity, enabling observation of a greater fraction of the metabolome. Relatively few MS-based fingerprinting studies have been reported to date, however, as several technical issues are still being addressed. These include whether there is a need for chromatographic separation prior to MS analysis, the degree of consistency between repeated analyses, and difficulties with metabolite quantification. Therefore, MS-based fingerprinting requires some further development and validation before implementation into a tiered risk assessment paradigm, and it would likely follow the NMR-based approach. Nevertheless, with the required support, MS methodologies could be achieved relatively quickly (within a few years).

The "black box" fingerprinting approach, however, must be treated with caution and, although such methods could add value to Tier 1 risk assessment in the near future, a more definitive and robust assessment of risk would be mechanistically based, which requires unambiguous identification of the transcripts and/or metabolites under study (profiling). In the case of transcriptomics, this requires that sequence annotation of the surrogate species is known. For some of the relevant species, this is available and these developments for application into tiered testing can commence here. There are a number of key organisms, however, for which the full genome sequence is urgently required (see Section 3.9).

Metabolic profiling has the advantage that no sequence information is required for the organism under investigation. This approach is therefore directly applicable to all organisms. However, the annotation of the many hundreds or thousands of peaks in NMR and mass spectra is currently a major challenge, especially for less well characterized nonmammalian species where completely novel metabolites will occur (Bundy et al. 2002). The creation of NMR spectral libraries of the more common metabolites is now underway both in industry (e.g., by Chenomx Inc. and

Bruker BioSpin) and by a number of initiatives in the public sector (e.g., the Human Metabolome Project [http://www.metabolomics.ca], The University of Birmingham, United Kingdom). Such endeavors will significantly increase the mechanistic information content of metabolite spectra and enable comprehensive analyses of particular pathways associated with particular mechanisms of toxicity. For example, targeted profiling of steroid pathways could include the observation of multiple steroid precursors (e.g., cholesterol, pregnenolone), sex steroids (e.g., testosterone, 11-ketotestosterone, 17β-estradiol, progesterone), and conjugated steroids (glucuronidated and sulphated). The close linkage between metabolic data and organism physiology suggests that metabolomics will be important for risk assessment by providing molecular correlates of organism survival, growth, and reproduction.

In summary, in the relatively short term (next 2 to 3 years), it is anticipated that transcriptome and metabolome fingerprinting could be implemented into tiered testing. To do so, however, easily accessible libraries of profile and fingerprinting data for a set of chemicals with well-defined, relevant MOA need to be developed. These fingerprints would then be used in a pattern matching approach to potentially identify the MOA of chemicals with previously unknown effects. Chemical selection and dosing regimes should be selected based upon the vast acute toxicity information that currently exists for the tiered testing species. Selection of substances for testing with known modes of action and known pathologies would ensure representation of important chemical classes and the development of a robust predictive model.

Additionally, development of fingerprints and profiles of response for a selection of substances in a variety of test species would enhance our capabilities for interspecies extrapolations. In conjunction with these activities, "proof of principle" case studies should be conducted to demonstrate utility of genomic data to the risk assessment process. Some xenoestrogens and pharmaceuticals (that work via specific mechanisms or pathways) would be particularly amenable for initiating studies into toxicogenomic fingerprinting and profiling. It is likely that only through demonstrated successes, such as those enabled through focused case studies, will the regulatory community accept toxicogenomics and commit additional resources.

For the integration of genomics into tiered testing, clear definitions of the "triggers" for directing higher tiers of testing need to be established. It is likely that almost all substances, both natural and synthetic, will induce gene expression and subsequent responses if present in sufficiently high concentrations, and a challenge will be to identify the critical fingerprints for the various classes of chemical substances of concern and thresholds of response that signal for concern in profiling. Triggers for fingerprinting can theoretically be characterized relatively quickly against a series of reference chemicals with defined MOAs and toxicity. Indeed, this is a strategy that is already in use for mammalian toxicity testing during drug development (Foreman 2007). Triggers for profiling (including NOTELs that might be applied to avoid higher tier testing) will be restricted to compounds where it is possible to predict the (highest) environmental concentrations with a good safety margin.

If genomics can efficiently and accurately predict long-term adverse effects from short-term transcript, protein, and metabolite profiles, then this technology has the potential to improve radically efficiency of today's hazard and risk assessment procedures in terms of speed, costs, vertebrate animal usage, and workability. A major medium- to long-term research objective and need involves studies focused on "phenotypic anchoring"—specifically, relating the molecular and biochemical responses measured by toxicogenomic methods to (adverse) alterations in the endpoints typically used for ecological risk assessments: survival, growth, and reproduction (Miracle and Ankley 2005). Characterizing the adverse phenotypes that result from classes of chemicals with well-established mechanisms of action could then potentially be used to guide the classification of the mechanisms of action of new chemicals.

Studies focused on phenotypic anchoring will involve additional costs and animals, as it is necessary to consider integrated responses across multiple biological levels of organization. However, this type of information is critical to supporting use of molecular data in regulatory programs. Phenotypic anchoring will also serve further to help in the understanding of compensatory (indirect) as opposed to direct effects of chemicals on the genome. Additionally, toxicogenomic data from phenotypic anchoring studies will be valuable in serving to provide meaningful linkages between single biomarkers (established or, indeed, new) and adverse outcomes.

In order to identify genes, metabolites, and patterns that are not predictive of adverse effects, studies also need to be undertaken at doses and concentrations that do not give rise to adverse outcomes, but where adaptive or compensatory molecular responses still may be detected. To further distill the normal variability in responses to fluctuations in environmental conditions not related to chemical exposure, studies on responses to changes in variables such as temperature, light, food availability, pH, parasitism, etc. should be pursued in relevant species in the mid- to longer term. Although such environmental variables most often can be controlled to a great extent in the laboratory, this type of background knowledge would greatly enhance the interpretation of any field or mesocosm studies at higher tiers.

As frequently emphasized throughout this chapter, a major incentive for improving the environmental risk assessment procedures is to refine and reduce vertebrate animal use. Genomic analyses combined with in vitro assays potentially have the capacity to predict modes of action and thus directly guide more specific tests at higher tiers. This would also be of significant value and it is therefore suggested that in vitro studies be performed in coordination with in vivo studies whenever possible in the development of the genomic or tiered testing database.

3.8 BIOINFORMATICS

In both the short- and longer term development of genomics in tiered testing, bioinformatics is a significant concern. Existing computational methods will be challenged by the very large quantity of data to be captured in an appropriate format from which it can be evaluated in an efficient and consistent manner. It is essential that the data capture and retrieval systems are standardized. Such approaches for data capture have already been initiated for transcriptomics in the United Kingdom and specifically for toxicological studies (e.g., minimum information about a microarray experiment [MIAME]-Tox) and environmental toxicology (e.g., MIAME-Env). In these formats, array data are logged onto a specified template, and all the relevant associated information on experimental design and biological endpoint effects (so-called metadata) are simultaneously captured. It is now a requirement for publication in the higher quality research journals that array data generation and analyses conform to these minimal standards. Parallel systems for the capture, retrieval, and interrogation of metabolomic data are also emerging (Jenkins et al. 2004; Lindon et al. 2005; Rubtsov et al. 2007). Such system requirements for qualified bioinformatics support and expertise are very significant and should not be underestimated in developing and applying the genomic techniques in tiered testing.

3.9 REGULATORY CHALLENGES AND RECOMMENDATIONS

The challenges for the development and application of genomics into the regulatory framework for tiered testing are not trivial. In the first instance, it requires the education of experts involved in risk assessments as to what toxicogenomics can (and cannot) be expected to accomplish. There is also the need for a standardization of terminology in genomics and for a simplification of jargon to make the science "more reachable" to all concerned. As a simple example on standardization of terminology, we propose that genomics-based fingerprinting be used in the context only to describe the unbiased approach to measure as many endpoints as possible where these endpoints are not identified; profile, signature, and pattern should be applied to approaches in which the endpoints are unambiguously identified.

A major challenge is the justification of increased resource investment for research and implementation of genomics into tiered testing in the immediate term with a potential "pay off" occurring some years in the future. In essence, this will mean an urgent need for proof of principle for fingerprinting, profiling, and assessments of the strengths of associations between phenotypic relationships and molecular responses (in both the short and longer term), established with chemicals with known MOAs. If successful, it will then be necessary to develop enhanced capacity for these tools and techniques and, indeed, expertise for widespread implementation into chemical testing.

Genomic research, as applied to human health studies and human toxicology, is generally more advanced than for ecotoxicology, and there needs to be careful consideration of the "lessons learned" for ecotoxicogenomics. Not least, there is the urgent need for a standardization of approaches, including of arrays; platforms; and approaches used in the capture, handling, and storage of data, as well as readily transferable guidance for the interpretation of the toxicogenomic data. A strong recommendation is that the genomics data capture systems should be harmonized across the OECD and USEPA. This needs to be implemented at a very early stage to avoid the loss of potentially valuable information as capacity is built.

Finally, there needs to be wide acceptance that paradigm shifts may be needed in the approach to tiered testing as genomics are developed to their full potential. This, in turn, means a re-evaluation of the approaches presently used for testing chemicals as is being explored via ITS.

REFERENCES

Aebersold R, Mann M. 2003. Mass spectrometry-based proteomics. Nature 422:198–207.

Albertsson E, Kling P, Gunnarsson L, Larsson DGJ, Forlin L. 2007. Proteomic analyses indicate induction of hepatic carbonyl reductase/20β-hydroxysteroid dehydrogenase B in rainbow trout exposed to sewage effluent. Ecotoxicol Environ Safety 68:33–39.

Allanou R, Hansen BG, van der Bilt Y. 1999. Public availability of data on EU high production volume chemicals; EUR 18996 EN. European Chemicals Bureau Web site. Http://ecb.jrc.it. Accessed 13 November 2006.

Archacki S, Wang Q. 2004. Expression profiling of cardiovascular disease. Hum Genomics 1:355–370.

Babin M, Casado S, Chana A, Herradon B, Segner H, Tarazona JV, Navas JM. 2005. Cytochrome P4501A induction caused by the imidazole derivative Prochloraz in a rainbow trout cell line. Toxicol in Vitro 19:899–902.

Baldessari D, Shin Y, Krebs O, Konig R, Koide T, Vinayagam A, Fenger U, Mochii M, Terasaka C, Kitayama A, et al. 2005. Global gene expression profiling and cluster analysis in *Xenopus laevis*. Mechan Devel 122:441–475.

Bapiro TE, Andersson TB, Otter C, Hasler JA, Masimirembwa CM. 2002. Cytochrome P-450 1A1/2 induction by antiparasitic drugs: dose-dependent increase in ethoxyresorufin O-deethylase activity and mRNA caused by quinine, primaquine and albendazole in HepG2 cells. Eur J Clin Pharmacol 58:537–542.

Bowman CJ, Kroll KJ, Hemmer MJ, Folmar LC, Denslow ND. 2000. Estrogen-induced vitellogenin mRNA and protein in sheepshead minnow (*Cyprinodon variegates*). Gen Compar Endocrinol 120:300–313.

Bradbury SP, Feijtel TCJ, van Leeuwen CJ. 2004. Meeting the scientific needs of ecological risk assessment in a regulatory context. Environ Sci Technol 38:463A–470A.

Bradley BP, Olsson B, Brown DC, Tedengren M. 1998. HSP70 levels in physiologically stressed Baltic Sea mussels. Marine Environ Res 46:397–400.

Brindle JT, Antti H, Holmes E, Tranter G, Nicholson JK, Bethell HWL, Clarke S, Schofield PM, McKilligin E, Mosedale DE, et al. 2002. Rapid and noninvasive diagnosis of the presence and severity of coronary heart disease using H-1 NMR-based metabonomics. Nature Med 8:1439–1444.

Bundy JG, Lenz EM, Bailey NJ, Gavaghan CL, Svendsen C, Spurgeon D, Hankard PK, Osborn D, Weeks JM, Trauger SA. 2002. Metabonomic assessment of toxicity of 4-fluoroaniline, 3,5-difluoroaniline and 2-fluoro-4-methylaniline to the earthworm *Eisenia veneta* (Rosa): identification of new endogenous biomarkers. Environ Toxicol Chem 21:1966–1972.

Bundy JG, Spurgeon DJ, Svendsen C, Hankard PK, Weeks JM, Osborn D, Lindon JC, Nicholson JK. 2004. Environmental metabonomics: applying combination biomarker analysis in earthworms at a metal contaminated site. Ecotoxicology 13:797–806.

Burkhardt-Holm P, Bernet D, Hogstrand C. 1999. Increase of metallothionein-immunopositive chloride cells in the gills of brown trout and rainbow trout after exposure to sewage treatment plant effluents. Histochem J 31:339–346.

Burns MA, He WL, Wu CL, Cheng LL. 2004. Quantitative pathology in tissue MR spectroscopy based human prostate metabolomics. Technol Cancer Res Treatment 3:591–598.

Cairns J, Dickson KL, Maki AW. 1979. Estimating the hazard of chemical substances to aquatic life. Hydrobiologia 64:157–166.

[CEPA] Canadian Environmental Protection Act. 1999. New Substances and Existing Substances Program and Domestic Substances List Categorization and Screening Program. Government of Canada Web site. www.ec.gc.ca/substances/index_e.html. Accessed 13 November 2006.

Celius T, Matthews JB, Giesy JP, Zacharewski TR. 2000. Quantification of rainbow trout (*Oncorhynchus mykiss*) zona radiate and vitellogenin mRNA levels using real-time PCR after in vivo treatment with estradiol-17 beta or alpha-zearalenol. J Steroid Biochem Molec Biol 75:109–119.

Celius T, Walther BT. 1998. Differential sensitivity of zonagenesis and vitellogenesis in Atlantic salmon (*Salmo salar* L.) to DDT pesticides. J Exp Zool 281:346–353.

ChemRTK (Chemical Right-to-Know) Initiative, The. 1998. USEPA's HPV Voluntary Children's Chemical Evaluation Program (VCCEP); April 21 1998; Environmental Protection Agency Web site. www.epa.gov/chemrtk/volchall.htm. Accessed 13 November 2006.

Chung-Davidson YW, Rees CB, Wu H, Yun SS, Li WM. 2004. Beta-naphthoflavone induction of CYP1A in brain of juvenile lake trout (*Salvelinus namaycush* Walbaum). J Exp Biol 207:1533–1542.

Combes R, Barratt M, Balls M. 2003. An overall strategy for the testing of chemicals for human hazard and risk assessment under the EU REACH system. ATLA 31:7–19.

[CEC] Commission of the European Communities. 1994. Commission regulation (EC) no. 1488/94 of 28 June 1994 laying down the principles for the assessment of risks to man and the environment of existing substances in accordance with Council Regulation (EEC) no. 793/93. Official Journal of the European Communities L161/3. Secretariat for European Affairs Web site. http://www.sei.gov.mk/npalreports/TU/en/31994R1488.doc. Accessed 13 November 2006.

Custodia-Lora N, Novillo A, Callard IP. 2004. Regulation of hepatic progesterone and estrogen receptors in the female turtle, *Chrysemys picta*: a relationship to vitellogenesis. Gen Comp Endocrinol 136:232–240.

Dunn WB, Ellis DL. 2005. Metabolomics: current analytical platforms and methodologies. Trends Anal Chem 24:285–294.

[EPA] Environmental Protection Agency. 1998. Chemical Hazard Data Availability Study. Washington, DC: US Government Printing Office. EPA Web site. www.epa.gov/chemrtk/hazchem.htm. Accessed 13 November 2006.

[EMEA] European Medicines Agency. 2005. Guidelines on the environmental risk assessment of pharmaceutical products for human use. EMEA Web site. www.emea.eu.int/pdfs/human/swp/444700en.pdf. Accessed 13 November 2006.

Fentem J, Chamberlain M, Sangster B. 2004. The feasibility of replacing animal testing for assessing consumer safety: a suggested future direction. ATLA 32:617–623.

Fisher MA, Mehne C, Means JC, Ide CF. 2006. Induction of CYP1A mRNA in carp (*Cyprinus carpio*) from the Kalamazoo River polychlorinated biphenyl-contaminated superfund site and in a laboratory study. Arch Environ Contamin Toxicol 50:14–22.

Foreman P. 2007. Toxicogenomics impacts decision-making early in the pipeline at Eli Lilly. Affymetrix Microarray Bulletin Web site. http://microarraybulletin.com/community/article.php?p=307. Accessed 16 April 2007.

Garcia-Reyero N, Raldua D, Quiros L, Llaveria G, Cerda J, Barcelo D, Grimalt JO, Pina B. 2004. Use of vitellogenin mRNA as a biomarker for endocrine disruption in feral and cultured fish. Anal Bioanal Chem 378:670–675.

George S, Burgess D, Leaver M, Frerichs N. 1992. Metallothionein induction in cultured fibroblasts and liver of a marine flatfish, the turbot, *Scophthalmus maximus*. Fish Physiol Biochem 10:43–54.

Gibb JOT, Holmes E, Nicholson JK, Weeks JM. 1997. Proton NMR spectroscopic studies on tissue extracts of invertebrate species with pollution indicator potential. Comp Biochem Physiol B 118:587–598.

Goksoyr A, Forlin L. 1992. The cytochrome p-450 system in fish, aquatic toxicology and environmental monitoring. Aquatic Toxicol 22:287–311.

Gracey AY, Fraser EJ, Li WZ, Fang YX, Taylor RR, Rogers J, Brass A, Cossins AR. 2004. Coping with cold: an integrative, multitissue analysis of the transcriptome of a poikilothermic vertebrate. Proc Natl Acad Sci USA 101:16970–16975.

Griffin JL, Walker LA, Garrod S, Holmes E, Shore RF, Nicholson JK. 2000. NMR spectroscopy based metabonomic studies on the comparative biochemistry of the kidney and urine of the bank vole (*Clethrionomys glareolus*), wood mouse (*Apodemus sylvaticus*), white toothed shrew (*Crocidura suaveolens*) and the laboratory rat. Comp Biochem Physiol B 127:357–367.

Griffin JL, Walker LA, Shore RF, Nicholson JK. 2001. High-resolution magic angle spinning H-1-NMR spectroscopy studies on the renal biochemistry in the bank vole (*Clethrionomys glareolus*) and the effects of arsenic (As3+) toxicity. Xenobiotica 31:377–385.

Gye MC, Kim DH. 2005. Bisphenol A induces hepatic vitellogenin mRNA in male *Bombina orientalis*. Bull Environ Contam Toxicol 75:1–6.

Hahn ME, Stegeman JJ. 1994. Regulation of cytochrome P450 1A1 in teleosts: sustained induction of CYP1A1 messenger-RNA, protein, and catalytic activity by 2,3,7,8 tetrachlorodibenzofuran in the marine fish *Stenotomus chrysops*. Toxicol Appl Pharmacol 127:187–198.

Hansen BG, van Haelst AG, van Leeuwen K, van der Zandt P. 1999. Priority setting for existing chemicals: European Union risk ranking method. Environ Toxicol Chem 18:772–779.

Hansson O, Zetterberg H, Buchhave P, Londos E, Blennow K, Minthon L. 2006. Association between CSF biomarkers and incipient Alzheimer's disease in patients with mild cognitive impairment: a follow-up study. Lancet Neurol 5:228–234.

Health Council of the Netherlands. 2001. Toxicity testing: a more efficient approach. Publication no. 2001/24E. The Hague (The Netherlands): Health Council of The Netherlands. 70 p.

Hemmer MJ, Bowman CJ, Hemmer BL, Friedman SD, Marcovich D, Kroll KJ, Denslow ND. 2002. Vitellogenin mRNA regulation and plasma clearance in male sheepshead minnows (*Cyprinodon variegates*) after cessation of exposure to 17 beta-estradiol and p-nonylphenol. Aquatic Toxicol 58:99–112.

Herman RL, Kincaid HL. 1988. Pathological effects of orally administered estradiol to rainbow trout. Aquaculture 72:165–172.

Hofer T, Gerner I, Gundert-Remy U, Liebsch M, Schulte A, Spielmann H, Vogel R, Wettig K. 2004. Animal testing and alternative approaches for the human health risk assessment under the proposed new European chemicals regulation. Arch Toxicol 78:549–564.

Huang YW, Matthews JB, Fertuck KC, Zacharewski TR. 2005. Use of *Xenopus laevis* as a model for investigating in vitro and in vivo endocrine disruption in amphibians. Environ Toxicol Chem 24:2002–2009.

Hutchinson TH, Ankley GT, Segner H, Tyler CR. 2006. Screening and testing for endocrine disruption in fish—biomarkers as signposts not traffic lights in risk assessment. Environ Health Perspect 114(Suppl 1):106–114.

Jenkins H, Hardy N, Beckmann M, Draper J, Smith AR, Taylor J, Fiehn O, Goodacre R, Bino RJ, Hall R, et al. 2004. A proposed framework for the description of plant metabolomics experiments and their results. Nat Biotechnol 22:1601–1606.

Kille P, Kay J, Leaver M, George S. 1992. Induction of piscine metallothionein as a primary response to heavy-metal pollutants—applicability of new sensitive molecular probes. Aquatic Toxicol 22:279–286

Kimerle RA, Levinskas GJ, Metcalf JS, Scharpf LG. 1977. An industrial approach to evaluating environmental safety of new products. In: Mayer FL, Hamelink JL, editors. Aquatic toxicology and hazard evaluation. Proceedings of the First Annual Symposium on Aquatic Toxicology. Philadelphia (PA): American Society for Testing Materials p. 36–43.

Knoebl I, Hemmer MJ, Denslow ND. 2004. Induction of zona radiata and vitellogenin genes in estradiol and nonylphenol exposed male sheepshead minnows (*Cyprinodon variegates*). Marine Environ Res 58:547–551.

Kramer VJ, Miles-Richardson S, Pierrens SL, Giesy JP. 1998. Reproductive impairment and induction of alkaline-labile phosphate, a biomarker of estrogen exposure, in fathead minnows (*Pimephales promelas*) exposed to 17β-estradiol. Aquatic Toxicol 40:335–360.

Krasnov A, Koskinen H, Pehkonen P, Rexroad CE, Afanasyev S, Molsa H. 2005. Gene expression in the brain and kidney of rainbow trout in response to handling stress. BMC Genomics 6:3.

Kucharski R, Maleszka R. 2005. Microarray and real-time PCR analyses of gene expression in the honeybee brain following caffeine treatment. J Molec Neurosci 27:269–276.

Lam KL, Ko PW, Wong JKY, Chan KM. 1998. Metal toxicity and metallothionein gene expression studies in common carp and tilapia. Marine Environ Res 46:563–566.

Larsson DGJ, Hyllner SJ, Haux C. 1994. Induction of vitelline envelope proteins by estradiol-17β in ten teleost species. Gen Comp Endocrinol 96:445–450.

Lee C, Na JG, Lee KC, Park K. 2002. Choriogenin mRNA induction in male medaka, *Oryzias latipes*, as a biomarker of endocrine disruption. Aquatic Toxicol 61:233–241.

Lettieri T. 2006. Recent applications of DNA microarray technology to toxicology and ecotoxicology. Environ Health Perspect 114:4–9.

Lindon JC, Nicholson JK, Holmes E, Antti H, Bollard ME, Keun H, Beckonert O, Ebbels TM, Reilly MD, Robertson D, et al. 2003. Contemporary issues in toxicology—the role of metabonomics in toxicology and its evaluation by the COMET project. Toxicol Appl Pharmacol 187:137–146.

Lindon JC, Nicholson JK, Holmes E, Everett JR. 2000. Metabonomics: metabolic processes studied by NMR spectroscopy of biofluids. Concepts Magnet Resonance 12:289–320.

Lindon JC, Nicholson JK, Holmes E, Keun HC, Craig A, Pearce JTM, Bruce SJ, Hardy N, Sansone SA, Antti H, et al. 2005. Summary recommendations for standardization and reporting of metabolic analyses. Nat Biotechnol 23:833–838.

Liney KE, Hagger JA, Tyler CR, Depledge MH, Galloway TS, Jobling S. 2006. Health effects in fish of long-term exposure to effluents from wastewater treatment works. Environ Sci Technol 114(Suppl 1):81–89.

Livingstone DR. 1993. Biotechnology and pollution monitoring—use of molecular biomarkers in the aquatic environment. J Chem Technol Biotechnol 57:195–211.

Lorenzen A, Williams KL, Moon TW. 2003. Determination of the estrogenic and antiestrogenic effects of environmental contaminants in chicken embryo hepatocyte cultures by quantitative-polymerase chain reaction. Environ Toxicol Chem 22:2329–2336.

Mathavan S, Lee SGP, Mak A, Miller LD, Murthy KRK, Govindarajan KR, Tong Y, Wu YL, Lam SH, Yang H, et al. 2005. Transcriptome analysis of zebrafish embryogenesis using microarrays. PLOS Genet 1:260–276.

Mattsson JL, Eisenbrandt DL, Doe JE. 2003. More than 10 000 animals are required for the registration of a single pesticide—this paradigm must be changed. Poster presented at 42nd Annual Meeting of the Society of Toxicology, Salt Lake City, Utah, 9–13 March 2003.

McClain JS, Oris JT, Burton GA, Lattier D. 2003. Laboratory and field validation of multiple molecular biomarkers of contaminant exposure in rainbow trout (*Oncorynchus mykiss*). Environ Toxicol Chem 22:361–370.

Miller DH, Jensen KM, Villeneuve DL, Kahl MD, Mallyen EA, Durhan EJ, Ankley GT. 2007. Linkage of biochemical responses to population-level effects. A case study with vitellogenin in the fathead minnow (*Pimephales promelas*). Environ Toxicol Chem 26:521–527.

Miracle AL, Ankley GT. 2005. Ecotoxicogenomics: linkages between exposure and effects in assessing risks of aquatic contaminants to fish. Reproduct Toxicol 19:321–326.

[NRC] National Research Council, Steering Committee on Identification of Toxic and Potentially Toxic Chemicals for Consideration by the National Toxicology Program. 1984. Toxicity testing strategies to determine needs and priorities. Washington, DC: National Academy Press, 396 p.

Nicholson JK, Connelly J, Lindon JC, Holmes E. 2002. Metabonomics: a platform for studying drug toxicity and gene function. Nature Rev Drug Discov 1:153–161.

Odunsi K, Wollman RM, Ambrosone CB, Hutson A, McCann SE, Tammela J, Geisler JP, Miller G, Sellers T, Cliby W, et al. 2005. Detection of epithelial ovarian cancer using H-1-NMR-based metabonomics. Int J Cancer 113:782–788.

[OECD] Organization for Economic Co-operation and Development. 1992. Report of the OECD workshop on the extrapolation of laboratory aquatic toxicity data to the real environment. OECD Environment Monographs No. 59. OECD Web site. http://www.oecd.org/dataoecd/30/48/34528236.pdf. Accessed 13 November 2006.

Pedersen F, De Bruijn J, Munn S, Van Leeuwen K. 2003. Assessment of additional testing needs under REACH. Effects of QSARs, risk based testing and voluntary industry initiatives. Report EUR 20863 EN, 33 p. Ispra (Italy): European Commission, Joint Research Centre. European Chemicals Bureau Web site. http://ecb.jrc.it. Accessed 13 November 2006.

Peiro G, Alary J, Cravedi JP, Rathahao E, Steghens JP, Gueraud F. 2005. Dihydroxynonene mercapturic acid, a urinary metabolite of 4-hydroxynonenal, as a biomarker of lipid peroxidation. Biofactors 24:89–96.

Rees CB, McCormick SD, Vanden Heuvel JP, Li WM. 2003. Quantitative PCR analysis of CYP1A induction in Atlantic salmon (*Salmo salar*). Aquatic Toxicol 62:67–79.

Reiner A, Yekutieli D, Benjamini Y. 2003. Identifying differentially expressed genes using false discovery rate controlling procedures. Bioinformatics 19:368–375.

Roberts AP, Cris JT, Burton GA, Clements WH. 2005. Gene expression in caged fish as a first-tier indicator of contaminant exposure in streams. Environ Toxicol Chem 24:3092–3098.

Robertson DG. 2005. Metabonomics in toxicology: a review. Toxicol Sci 85:809–822.

Rosenblum ES, Viant MR, Braid BM, Moore JD, Friedman CS, Tjeerdema RS. 2005. Characterizing the metabolic actions of natural stresses in the California red abalone, *Haliotis rufescens*, using ^{1}H NMR metabolomics. Metabolomics 1:199–209.

Rubtsov DV, Jenkins H, Ludwig C, Easton J, Viant MR, Guenther UL, Griffin JL, Hardy N. 2007. Proposed reporting requirements for the description of NMR-based metabolomics experiments. Metabolomics. Forthcoming.

Ruetschi U, Zetterberg H, Podust VN, Gottfries J, Li S, Hviid Simonsen A, McGuire J, Karlsson M, Rymo L, Davies H, et al. 2005. Identification of CSF biomarkers for frontotemporal dementia using SELDI-TOF. Exp Neurol 196:273–281.

Russell WMS, Burch RL. 1959. The principles of humane experimental technique. London: Methuen. 238 p.

Samuelsson L, Forlin L, Karlsson G, Adolfsson-Erici M, Larsson DGJ. 2006. Using NMR metabolomics to identify responses of an environmental estrogen in blood plasma of fish. Aquatic Toxicol 78:341–349.

Santos EM, Workman V, Paull GC, Filby AL, Van Look KW, Kille P, Tyler CR. 2007. Molecular basis of sex and reproductive status in breeding zebrafish. Physiol Genomics Forthcoming.

Schmid T, Gonzalez-Valero J, Rufli H, Dietrich DR. 2002. Determination of vitellogenin kinetics in male fathead minnows (*Pimephales promelas*). Toxicol Lett 131:65–74.

Schmidt CW. 2004. Metabolomics: What's happening downstream of DNA. Environ Health Perspect 112:A410–A415.

Schrader EA, Henry TR, Greeley MS Jr, Bradley BP. 2003. Proteomics in zebrafish exposed to endocrine disrupting chemicals. Ecotoxicology 12:485–488.

Schulz RW, Boegard J, Male R, Ball J, Fenske M, Olsson L, Tyler CR. 2007. Estrogen-induced alterations in amh and dmrt1 expression signal for disruption in male sexual development in the zebrafish. Environ Sci Technol. Forthcoming.

Schwaiger J, Spieser OH, Bauer C, Ferling H, Mallow U, Kalbfus W, Negele RD. 2000. Chronic toxicity of nonylphenol and ethinylestradiol: haematological and histopathological effects in juvenile common carp (*Cyprinus carpio*). Aquatic Toxicol 51:69–78.

Soetaert A, Moens LN, Van der Ven K, Van Leemput K, Naudts B, Blust R, De Coen WM. 2006. Molecular impact of propiconazole on *Daphnia magna* using a reproduction-related cDNA array. Comp Biochem Physiol C 142:66–76.

Solanky KS, Burton IW, MacKinnon SL, Walter JA, Dacanay A. 2005. Metabolic changes in Atlantic salmon exposed to *Aeromonas salmonicida* detected by H-1-nuclear magnetic resonance spectroscopy of plasma. Diseases Aquatic Organ 65:107–114.

Stentiford GD, Viant MR, Ward DG, Johnson PJ, Martin A, Wenbin W, Cooper HJ, Lyons BP, Feist SW. 2005. Liver tumors in wild flatfish: a histopathological, proteomic, and metabolomic study. OMICS 9:281–299.

Sumpter JP, Jobling S. 1995. Vitellogenesis as a biomarker for estrogenic contamination of the aquatic environment. Environ Health Perspect 103:173–178.

Thebault MT, Raffin JP, Picado AM, Mendonca E, Skorkowski EF, Le Gal Y. 2000. Coordinated changes of adenylate energy charge and ATP/ADP: use in ecotoxicological studies. Ecotoxicol Environ Safety 46:23–28.

Thomas CE, Ganji G. 2006. Integration of genomic and metabonomic data in systems biology—are we "there" yet? Curr Opin Drug Discov Devel 9:92–100.

Thomas-Jones E, Thorpe K, Harrison N, Thomas G, Morris C, Hutchinson T, Woodhead S, Tyler C. 2003. Dynamics of estrogen biomarker responses in rainbow trout exposed to 17 beta-estradiol and 17 alpha-ethinylestradiol. Environ Toxicol Chem 22:3001–3008.

Thomas-Jones E, Walkley N, Morris C, Kille P, Cryer J, Weeks I, Woodhead JS. 2003. Quantitative measurement of fathead minnow vitellogenin mRNA using hybridization protection assays. Environ Toxicol Chem 22:992–995.

Thorpe KL, Tyler CR. 2006. Oestrogenic endocrine disruption in fish—developing biological effect measurement tools and generating hazard data. R&D Technical Report. Bristol (UK): Environment Agency (SC 00043/SR). 91 p.

Tom M, Chen N, Segev M, Herut B, Rinkevich B. 2004. Quantifying fish metallothionein transcript by real time PCR for its utilization as an environmental biomarker. Marine Pollut Bull 48:705–710.

[TSCA] The Toxic Substances Control Act. 1976. US Code, Title 15—Commerce and Trade, chapter 53, subchapter 1, section 2601—findings, policy and intent. US Code Web site. http://www.access.gpo.gov/uscode/title15/chapter53_.html. Accessed 13 November 2006.

Vaccaro E, Meucci V, Intorre L, Soldani G, Di Bello D, Longo V, Gervasi PG, Pretti C. 2005. Effects of 17 beta-estradiol, 4-nonylphenol and PCR 126 on the estrogenic activity and phase 1 and 2 biotransformation enzymes in male sea bass (*Dicentrarchus labrax*). Aquatic Toxicol 75:293–305.

Van der Jagt K, Munn S, Tørsløv J, De Bruijn J. 2004. Alternative approaches can reduce the use of test animals under REACH. Addendum to the report "Assessment of additional testing needs under REACH. Effects of (Q)SARs, risk based testing and voluntary industry initiatives". Report EUR 21405. Ispra (Italy): European Commission, Joint Research Centre. European Chemicals Bureau Web site. http://ecb.jrc.it/DOCUMENTS/REACH/PUBLICATIONS/Reducing_the_use_of_test_animals_under_REACH_IHCP_report.pdf. Accessed 13 November 2006.

Van Hemert S, Hoekman AJ, Smits MA, Rebel JMJ. 2004. Differences in intestinal gene expression profiles in broiler lines varying in susceptibility to malabsorption syndrome. Poultry Sci 83:1675–1682.

van Leeuwen CJ, BroRasmussen F, Feijtel TCJ, Arndt R, Bussian BM, Calamari D, Glynn P, Grandy NJ, Hansen B, VanHemmen JJ, et al. 1996. Risk assessment and management of new and existing chemicals. Environ Toxicol Pharmacol 2:243–299.

van Leeuwen CJ, Hermens JLM, editors. 1995. Risk assessment of chemicals. An introduction. Dordrecht (the Netherlands): Kluwer Academic Publishers, 374 p.

van Leeuwen CJ, Patlewicz GY, Worth A. 2006. Intelligent testing strategies. In: van Leeuwen CJ, Vermeire T, editors. Risk assessment of chemicals. An introduction. 2nd ed. Dordrecht (the Netherlands): Springer. Forthcoming.

Verslycke T, Poelmans S, De Wasch K, De Brabander HF, Janssen CR. 2004. Testosterone and energy metabolism in the estuarine mysid *Neomysis integer* (Crustacea: Mysidacea) following exposure to endocrine disruptors. Environ Toxicol Chem 23:1289–1296.

Viant MR, Bundy JG, Pincetich CA, de Ropp JS, Tjeerdema RS. 2005. NMR-derived developmental metabolic trajectories: an approach for visualizing the toxic actions of trichloroethylene during embryogenesis. Metabolomics 1:149–158.

Viant MR, Pincetich CA, Hinton DE, Tjeerdema RS. 2006. Toxic actions of dinoseb in medaka (*Oryzias latipes*) embryos as determined by in vivo ^{31}P NMR, HPLC-UV and ^{1}H NMR metabolomics. Aquatic Toxicol 76:329–342.

Viant MR, Pincetich CA, Tjeerdema RS. 2006. Metabolic effects of dinoseb, diazinon and esfenvalerate in eyed eggs and alevins of Chinook salmon (*Oncorhynchus tshawytscha*) determined by ^{1}H NMR metabolomics. Aquatic Toxicol 77:359–371.

Viant MR, Rosenblum ES, Tjeerdema RS. 2003. Identification of biomarkers for withering syndrome in red abalone using NMR-based metabonomics. Toxicol Sci 72:1165.

Viant MR, Walton JH, TenBrook PL, Tjeerdema RS. 2002. Sublethal actions of copper in abalone (*Haliotis rufescens*) as characterized by in vivo ^{31}P NMR. Aquatic Toxicol 57:139–151.

Viant MR, Werner I, Rosenblum ES, Ganter AS, Tjeerdema RS, Johnson ML. 2003. Correlation between heat-shock protein induction and reduced metabolic condition in juvenile steelhead trout (*Oncorhynchus mykiss*) chronically exposed to elevated temperature. Fish Physiol Biochem 29:159–171.

Vornanen M, Hassinen M, Koskinen H, Krasnov A. 2005. Steady-state effects of temperature acclimation on the transcriptome of the rainbow trout heart. Am J Physiol—Regul Integr Compar Physiol 289:1177–1184.

Vrana KE, Freeman WM, Aschner M. 2003. Use of microarray technologies in toxicology research. Neurotoxicology 24:321–332.

Wadlow R, Ramaswamy S. 2005. DNA microarrays in clinical cancer research. Curr Molec Med 5:111–120.

Winter MJ, Verweij F, Garofalo E, Ceradini S, McKenzie DJ, Williams MA, Taylor EW, Butler PJ, van der Oost R, Chipman JK. 2005. Tissue levels and biomarkers of organic contaminants in feral and caged chub (*Leuciscus cephalus*) from rivers in the West Midlands, UK. Aquatic Toxicol 73:394–405.

Wong CKC, Yeung HY, Cheung RYH, Yung KKL, Wong MH. 2000. Ecotoxicological assessment of persistent organic and heavy metal contamination in Hong Kong coastal sediment. Arch Environ Contamin Toxicol 38:486–493.

Worth AP. 2004. The tiered approach to toxicity assessment based on the integrated use of alternative (non-animal) tests. In: Cronin MTD, Livingstone D, editors. Predicting chemical toxicity and fate. Boca Raton (FL): CRC Press. p. 389–410.

Worth AP, Fentem JH, Balls M, Botham PA, Curren RD, Earl LK, Esdaile DJ, Liebsch M. 1998. An evaluation of the proposed OECD testing strategy for skin corrosion. ATLA 26:709–720.

Yang J, Xu GW, Zheng YF, Kong HW, Pang T, Lv S, Yang Q. 2004. Diagnosis of liver cancer using HPLC-based metabonomics avoiding false-positive result from hepatitis and hepatocirrhosis diseases. J Chromatogr B—Anal Technol Biomed Life Sci 813:59–65.

Zaroogian G, Gardner G, Horowitz DB, Gutjahr-Gobell R, Haebler R, Mills L. 2001. Effect of 17 beta-estradiol, o,p′-DDT, octylphenol and p,p-′DDE on gonadal development and liver and kidney pathology in juvenile male summer flounder (*Paralichthys denatus*). Aquatic Toxicol 54:101–112.

4 Application of Genomics to Regulatory Ecological Risk Assessments for Pesticides

Sigmund J Degitz, Robert A Hoke, Steven Bradbury, Richard Brennan, Lee Ferguson, Rebecca Klaper, Laszlo Orban, David Spurgeon, and Susan Tilton

CONTENTS

4.1	Background and Regulatory Framework 64
	4.1.1 Priority Setting: Near to Midterm Regulatory and Scientific Challenges 65
	4.1.2 Evolution of a Hypothesis-Driven Risk Assessment Paradigm for Conventional Pesticides: Long-Term Regulatory and Scientific Challenges 65
	4.1.3 Developing a Coordinated Regulatory and Research Program 65
4.2	Incorporation of Genomic Technology in the Risk Assessment Paradigm 66
4.3	Pesticide Risk Assessment Process, Challenges, and Potential Applications for Genomics 68
	4.3.1 EPA, OPP, Ecological Risk Assessment (ERA) Process—Conventional Outdoor Pesticide Active Ingredients 68
	4.3.1.1 Application of Genomics to Exposure Assessment 70
	4.3.1.2 Application of Genomics to Effects Characterization 72
	4.3.1.3 Application of Genomics to Risk Characterization 75
4.4	Risk Assessment for Antimicrobials and Inert Ingredients 78
	4.4.1 Antimicrobials 78
	4.4.1.1 Current Practices 78
	4.4.2 Inert Ingredients 79
	4.4.2.1 Current Practices 79
	4.4.3 Application of Genomics to ERA for Inert Ingredients and Antimicrobials 80

4.5   Necessary Developments in Application of Genomic Approaches to
       Risk Assessment of Pesticides, Antimicrobials, and Inert Ingredients 80
       4.5.1 Establish Baseline Variation of the Transcriptome, Proteome,
              and Metabolome 80
       4.5.2 Cross-Species Experiments Using Existing Tools 81
       4.5.3 Development of Genomic Tools for the Most Ecologically
              Relevant Model Species 82
       4.5.4 Need for Large-Scale Demonstration Projects 83
References 83

4.1 BACKGROUND AND REGULATORY FRAMEWORK

Substantial advances in human health and ecological risk assessment have been achieved by the risk assessment community; however, challenges remain, such as providing credible scientific information on a timely, efficient basis to support decisions for industrial chemicals and pesticides. Current risk assessment data generation requirements—including animal welfare concerns and the volume, appropriateness, and cost of required data—and the large number of chemicals requiring evaluation are a challenge confronting the chemical industry, national and international regulatory agencies, and associated stakeholders (Bradbury et al. 2004). The lack of hazard data for many chemicals and the need to improve the efficiency and quality of risk assessment and management of chemicals are driving forces behind implementation of the Federal Insecticide, Fungicide and Rodenticide Act (FIFRA), the Food Quality Protection Act (FQPA), and European Union (EU) legislation being implemented under the auspices of the registration, evaluation, authorization, and restriction of chemicals (REACH) framework (http://ec.europa.eu/environment/chemicals/reach.htm).

Data are needed to address the risk assessment uncertainties arising from the lack of knowledge across chemical classes and adverse effects and outcomes of concern. However, the magnitude of the knowledge gap and the desire of regulatory agencies and affected stakeholders to close the knowledge gap as quickly as possible, preclude continued reliance solely on traditional toxicity testing for hazard evaluation. The long-term solution will not be the quicker generation of more data but rather the determination of which specific effects data, for which chemicals and exposures, are essential to assess and manage risks appropriately. The risk assessment process requires targeted, credible information for decision making—not an excess of information potentially unrelated to effects of concern. Such a process is also consistent with the efficient use of time and resources for the generation and interpretation of toxicity data, as well as the responsible use of animals for testing. A similar approach is needed to determine the types and amount of exposure data necessary to screen a chemical with a potentially wide range of uses. With this perspective as background, we propose an approach for advancing a sustainable evolution in the risk assessment and management paradigm that suggests how genomic data can help achieve these goals.

4.1.1 PRIORITY SETTING: NEAR TO MIDTERM REGULATORY AND SCIENTIFIC CHALLENGES

The challenge is efficient and effective prediction of hazard and exposure for chemicals that have limited data (e.g., pesticide inert ingredients, antimicrobial pesticides) in order to facilitate identification of the necessary empirical studies. Understanding of initiating events and the subsequent cascade of events leading to observed effects (i.e., toxicity pathways) will advance the development and use of genomic techniques to estimate hazard potential and support ranking and prioritization of chemicals for their potential to elicit adverse outcomes. Over the next 5 years, implementation of in silico and in vitro systems to advance prioritization and screening of chemicals such as inerts and antimicrobials will extend our methods for predicting toxicological potential and highlight the near- to mid-term research needs.

4.1.2 EVOLUTION OF A HYPOTHESIS-DRIVEN RISK ASSESSMENT PARADIGM FOR CONVENTIONAL PESTICIDES: LONG-TERM REGULATORY AND SCIENTIFIC CHALLENGES

Where data generation is required to make regulatory decisions (e.g., conventional outdoor use pesticides), the challenge is to develop the means to move, in a scientifically defensible and transparent manner, from a paradigm that requires extensive hazard and exposure testing, followed by the elimination of information irrelevant to the assessment, to a paradigm that provides a risk-based, hypothesis-driven approach to identify the specific in vivo information most relevant to the assessment. In this context, the fundamental paradigm shift is essentially the same for "data-rich" and "data-poor" chemicals. Understanding toxicity pathways from initiating events to adverse outcomes will advance development of testing designs that estimate in vivo potency for targeted endpoints, species, and life stages of greatest interest relative to the identified toxicity pathway. Achieving these scientific advances will facilitate identification of research strategies to advance, over the longer term, diagnostic approaches that are more efficient for assessing risks when in vivo data are necessary.

4.1.3 DEVELOPING A COORDINATED REGULATORY AND RESEARCH PROGRAM

A strategic direction and plan are needed to ensure critical research and regulatory needs are being met in an appropriate sequence and time frame and to allocate investment resources across various implementation efforts to achieve a holistic process. Partnerships among national and international governmental agencies and organizations (e.g., US Environmental Protection Agency [USEPA], Food and Drug Administration [FDA], National Institute of Environmental Health Sciences [NIEHS], Organization for Economic Cooperation and Development [OECD], European Union [EU]), as well as the regulated community, will be required to achieve this goal. There has been considerable progress in this area to date; for example, the European Commission recently published a perspective on "integrative testing" as a framework to advance development of in silico, in vitro, and efficient in vivo data sets (European Commission 2005).

In addition, the International Life Sciences Institute–Health and Environmental Sciences Institute (ILSI–HESI) program on the application of genomics to mechanism-based risk assessment is broadly addressing hepatotoxicity, nephrotoxicity, and genotoxicity. HESI has also partnered with the European Bioinformatics Institute to develop a database to house the data generated as part of its collaborative research. The HESI Technical Committee on Agricultural Chemical Safety Assessment (ACSA) is addressing how the current toxicology testing paradigm for pesticides could be improved to minimize the number of animals used while generating toxicity data that can be applied to assess a range of relevant pesticide exposure scenarios and provide a thorough and integrated evaluation across life stages (Barton et al. 2006; Carmichael et al. 2006; Cooper et al. 2006; Doe et al. 2006).

In addition, a committee of the National Research Council (NRC) is conducting a two-part study to assess current approaches to toxicity testing and assessment to meet regulatory data needs. The committee will consider evolving regulatory data needs, current toxicity testing guidelines, and emerging science and tools, as well as the challenge of incorporating more complex information (e.g., toxicokinetics, mechanisms of action, systems biology) into human health risk assessment. Finally, in the United Kingdom, the Royal Commission on Environmental Pollution (2003) made a number of recommendations concerning approaches to better risk assessment, including improvement in the application of monitoring methods and increasing efforts by industry and regulatory agencies to "augment genomics research significantly, in a direction that will lead towards an understanding of the way that synthetic chemicals interact with biological organisms". Considered together, these efforts provide evidence of the widespread recognition of the need for improvements in the risk assessment process and also demonstrate that preliminary efforts are underway among multiple groups to develop a coordinated and strategic vision to achieve this goal.

4.2 INCORPORATION OF GENOMIC TECHNOLOGY IN THE RISK ASSESSMENT PARADIGM

Development of data to facilitate comparisons across chemical classes and species for adverse effects and outcomes of concern is a critical need to improve the risk assessment process and help reduce uncertainties. Classical in vivo toxicology tests focus primarily upon tissue- and organism-level effects, which are often insufficient for discriminating mechanisms of action of chemicals. Data are required that relate chemical structure to specific mechanisms of action that can then be related to adverse outcomes. A fundamental understanding of mechanisms of action also is critical to the ability to extrapolate toxicological effects among species and chemicals and levels of biological organization.

Until recently, the major challenges have been in development of the capacity to generate these data. Several recent technological advances now make it possible to develop molecular profiles using transcriptomic, proteomic, and metabolomic methods. Conducting these types of analyses is no longer a question of capability. We envision that these measurements, taken in concert with one another, will become the standard approach for elucidating toxicological mechanisms and pathways.

The development of these technologies has led to the evolution of three scientific disciplines (see Chapter 1 for a more detailed description of these techniques):

genomics—the study of genes and their functions
proteomics—the study of the full set of proteins encoded by a genome
metabolomics—the study of the total metabolite pool

In toxicology, the integration of genomics and proteomic responses through the use of bioinformatic tools is termed toxicogenomics. When examination of the cellular response is extended to include metabolomics, a complete molecular profile can be developed that represents the mechanism of action for a compound. Each of these disciplines plays a role in molecular fingerprinting. Transcriptomics, a subset of genomics, is defined as the study of gene expression by measurement of mRNA. It is particularly useful because it facilitates the analysis of gene expression for most functional genes. Proteomics allows for the analysis of actual mRNA translation, as well as post-translational modifications (e.g., phosphorylation, glycosylation) that regulate protein function. Metabolomics essentially defines the metabolic and physiologic status of cells and tissues.

Transcript fingerprinting is a significant tool for drug safety evaluation in the pharmaceutical industry (Fielden, Eynon, et al. 2005; Fielden, Pearson, et al. 2005). The need for expression fingerprinting and reference data sets of chemically induced gene expression patterns for comparison with the effects of novel compounds has generated an economic opportunity. Several companies now provide high-quality, well-annotated, reproducible and experimentally tractable expression fingerprinting platforms; outsourcing of expression fingerprinting services; and databases of compound effects on gene expression in various systems with expert expression data analysis services.

The FDA has taken a lead regulatory role in promoting new molecular techniques to improve efficiency in the drug development process and has taken significant steps to encourage and prepare for the submission of genomic data as part of drug safety submissions (USFDA 2005). As awareness and acceptance increase in the regulatory community, DNA microarray data will be better integrated into chemical evaluation to facilitate identification and evaluation of problems at an earlier stage of the regulatory process. The FDA has published a guidance document (http://www.fda.gov/cder/guidance/6400fnl.pdf) with recommendations on when to submit genomic data during the product development process, the format to be used for data submission, and how the data will be used in regulatory decisions.

The establishment of a cross-center Interdisciplinary Pharmacogenomic Review Group (IPRG) to review voluntary genomic data submissions (VGDSs) encourages stakeholders to evaluate the role of genomic data in regulatory submissions by creating an opportunity for early informal meetings with the agency, providing informal peer-review feedback on genomic issues and/or questions, and offering insight into current FDA thinking about genomics. These interactions may assist in reaching important strategic decisions as well as providing an opportunity for sponsors to influence the FDA's thinking and help build consensus around future policies.

Major uses of toxicogenomic data in the drug development process include the mechanistic characterization of toxicity (Bulera et al. 2001; Steiner et al. 2004; Tugendreich et al. 2006), development of sensitive and specific biomarkers that are diagnostic and/or predictive of chemically induced adverse effects (Ruepp et al. 2005; Fielden, Eynon, et al. 2005; Fielden, Pearson, et al. 2005), prioritization of lead candidate compounds for development (Lindon et al. 2003; Ganter et al. 2005), and comparative analyses and ranking of compounds within chemical or pharmacological classes (Sawada et al. 2005; Yang et al. 2006). Many of the approaches and databases of expression information developed in the pharmaceutical industry are immediately applicable to the ecological risk assessment process for pesticides and industrial chemicals—particularly in the areas of identification of mechanism or mode of action, development of screening tests for toxicities of ecological concern, and efficient prioritization of compounds for targeted, in-depth evaluations. Additional research and development will be required before a broader application to ecological risk assessment is possible and before such data gain acceptance as a surrogate for existing regulatory requirements.

The EPA Office of Research and Development (ORD) research program on computational toxicology also is seeking to leverage new technologies, including genomics, proteomics, and metabolomics, to improve risk assessment. The strategic objectives of the EPA program are to develop improved linkages across the source-to-outcome continuum, approaches for prioritizing chemicals for subsequent screening and testing, and better methods and predictive models for quantitative risk assessment (http://www.epa.gov/comptox/).

4.3 PESTICIDE RISK ASSESSMENT PROCESS, CHALLENGES, AND POTENTIAL APPLICATIONS FOR GENOMICS

4.3.1 EPA, OPP, Ecological Risk Assessment (ERA) Process— Conventional Outdoor Pesticide Active Ingredients

Within the United States, EPA, through the Office of Pesticide Programs (OPP), is tasked to regulate and evaluate the use of all pesticides (including antimicrobials) and inert ingredients under FIFRA with the goal of protecting the health of humans and ecosystems from adverse effects related to pesticide exposure. In parallel, the European Union regulates pesticides via a similar process (http://www.efsa.europa.eu/en/about_efsa.html). There are two main processes in the regulation life cycle of pesticides within the United States: registration of new pesticides and re-registration of pesticides currently in use.

The risk assessment process used by OPP follows a broader ERA framework (USEPA 2004). Before the risk assessment process begins, risk assessors and risk managers discuss 1) the potential value of conducting a risk assessment; 2) goals for ecological resources; 3) range of management options; 4) objectives of the risk assessment; 5) the focus, scope, and timing of the assessment; and 6) resource availability. The characteristics of an ERA are directly determined by agreements reached by risk managers and assessors during early planning meetings.

The typical assessment endpoints for pesticide ERA are reduced survival and reproductive impairment for both aquatic and terrestrial animal species from both

TABLE 4.1
Typical Species Utilized for Toxicity Testing in Support of Pesticide Risk Assessment[a]

Birds (mallard duck and bobwhite quail)—surrogate for terrestrial-phase amphibians and reptiles

Mammals (laboratory rat)

Freshwater fish (bluegill sunfish, rainbow trout, and fathead minnow)—surrogate for aquatic phase amphibians

Freshwater invertebrates (water flea)

Estuarine and marine fish (sheepshead minnow)

Estuarine and marine invertebrates (oyster, mysid shrimp)

Terrestrial plants (corn, soybean, carrot [radish or sugar beet], oat [wheat or ryegrass], tomato, onion, cabbage [cauliflower or Brussels sprout], lettuce, cucumber)

Algae and aquatic plants (duckweed, several species of freshwater and marine algae)

[a] After Touart LW, Maciorowski AF. 1997. Ecol Appl 7(4):1086–1093.

---

direct acute or chronic exposures. Although they are measured at the individual level, these assessment endpoints provide insight about risks at higher levels of biological organization (e.g., populations). The EPA has developed regulations (40 CFR Part 158) that specify the types and amount of information required to support the registration of pesticide products. Rarely are toxicity data available for the actual receptor species identified in the risk assessment. In most cases, the ERA process relies on a suite of toxicity studies performed on a small number of surrogate organisms in the taxonomic groupings identified in Table 4.1.

Acute and chronic toxicity test endpoints are selected for the most sensitive species from the available test data within each of the taxonomic groups. If additional toxicity data are available for other species in a particular group, these data may be considered in the identification of the most sensitive species and endpoints. If additional data are available, decisions are made regarding the quality and utility of the data in the risk assessment (e.g., a review of the validity and reliability of study protocols).

Exposures estimated in the risk assessment for nontarget organisms are not specific to a given species. Aquatic organism (plant and animal) exposures are based on a set of standardized water body assumptions (water body size, watershed size, proximity to field, etc.) that result in overly conservative predictions of exposure. Estimates of exposure for terrestrial birds and mammals assume that animals are in the treatment area; they are typically based on taxa groupings based on food preferences (e.g., obligate insectivores, herbivores, granivores) and generic weight classes. Exposure for terrestrial plants considers surface runoff from treated fields as well as direct application via pesticide spray drift. Exposure estimates help to determine whether organisms in the environment will come in contact with the chemical. Estimating risk involves determining not only the toxicity of a compound but also whether organisms of interest in the aquatic and terrestrial environment will come into contact with the compound at concentrations likely to cause harm.

4.3.1.1 Application of Genomics to Exposure Assessment

Genomic data could provide a significant tool for examining exposure to a pesticide either in field trials prior to registration or from monitoring data collected after the chemical has been in use. These data also offer a new way to examine exposure by providing insight into some of the common issues related to exposure characterization, including:

- assessment of the effects of transient chemical exposures in field situations (i.e., whether low or no tissue residues in organisms equate to no exposure)
- the inability to determine potential exposures for some commensal uses
- information on whether certain organisms frequent areas where the pesticide is applied
- potential secondary (i.e., dietary) exposure
- potential for delayed effects

Gene expression, protein, and metabolite changes can indicate that an organism has been exposed to a chemical. For example, in environmental studies, certain biomarkers, proteins, or enzymes can indicate exposure to particular pollutants (Table 4.2). However, it is important to note that these biomarkers have not typically been used in the data dossiers used to support pesticide registrations due to the uncertainties or difficulties involved in linking most biomarker responses to higher level effects of interest (Forbes and Palmquist 2006). Genomic markers could provide an indication of exposure when it is otherwise difficult to determine exposure analytically (e.g., due to lack of an analytical method, insufficient analytical sensitivity, etc.), potentially providing an early indicator of exposure at concentrations lower than are currently detectable through instrumental techniques (Klaper and Thomas 2004). If gene expression, protein, or metabolite changes persist after the compound is no longer present in the organism or environment, these approaches could provide additional benefit by identifying potential exposure to chemicals causing persistent effects.

Various examples of genomic indicators of exposure exist and include detection of exposure to endocrine-disrupting compounds and mixtures of pollutants (Larkin et al. 2003; Denslow et al. 2004). Transcriptomics and proteomics may also provide ways to separate exposure to various compounds. If each class of compounds provides a unique expression signature, then expression patterns can indicate exposure to distinct chemical classes. Assuming chemicals affect a specific set of pathways, monitoring different points in the pathway of interest using gene expression can provide a sensitive indicator of exposure.

Single biomarkers provide an example of the utility of this approach. For example, changes in acetylcholinesterase activity are currently used to evaluate exposure to organophosphate and carbamate compounds (USEPA 2000). Genomic biomarkers provide the ability to monitor hundreds to thousands of genomic responses, thereby potentially providing a more specific signature for a given compound or class of compounds. There is some indication that even within a chemical class, genomic signatures may be specific (Hamadeh et al. 2002), thereby providing a means to

TABLE 4.2
Selected Biomarkers for Exposure to Various Types of Chemicals

Biomarker	Current Use
Aminolevulinic acid dehydratase (ALAD)	Metal exposure assessment, effects on heme production
Acetylcholinesterase (AChE)	Exposure to neurotoxins, organophosphates, carbamates, etc.
Cytochrome P450	Exposure to dioxins, trichloroethylene, other organic compounds, etc.
DNA adduct concentration	Exposure to chemical carcinogen
Hematology	Hemoglobin, hematocrit, plasma protein for general dysfunction (e.g., creatine phosphokinase, lactic dehydrogenase, glutamic-oxaloacetic)
Plasma enzyme activity	Transaminase, glutamic-pyruvic transaminase for stress in rats
Hormone	Ecdysone, testosterone, estrogen, etc., measure for altered reproductive function
Protoporphyrin	Exposure to metals
Metallothionien	Metal exposure
Alkaline phosphotase	Test for liver damage
Alanine aminotransferase	Test for liver damage
Egg shell thinning	Test for organochlorine exposure
Glutathione S-transferase (GST)	Measure of Phase II reaction, exposure to nitrocompounds, organophosphates, organochlorines
Lipid peroxidation	Measure of potential oxidate damage
Superoxide dismutase	Measure of antioxidant enzymes—exposure to polycyclic aromatic hydrocarbon (PAH), polychlorinated biphenyl (PCB), dioxin, organotin
DT-diaphorase	Measure of antioxidant enzymes—exposure to PAH, PCB, dioxin, organotin
HSPs	General stress—sublethal toxicity
EROD (ethoxyresorufin-O-deethylase)	Exposure to PAH, polyhalogenated hydrocarbon (PHH), evidence of receptor-mediated induction of cytochrome P450-dependent monooxygenases
Vitellogenin	Exposure to estrogenic substances
Choriogenin	Exposure to estrogenic substances
Nuclear receptor expression	Endocrine disruption
Chitinase	Exposure to substance that disrupts invertebrate molting
Protein adducts	Exposure to reactive organic compounds

determine exposure to single chemicals. Development of a "signature" database for specific classes of compounds would enable genomic markers to be used to identify exposure to a single chemical or class of chemicals that an organism may be exposed to in the environment. It will also be necessary, however, to understand the effects of environmental and other uncontrolled variables on these genomic fingerprints.

In order for the potential of genomics to be realized, data must be generated to determine whether genomic approaches provide improvements over existing exposure assessment methods. Genomic methods also need to be cost effective, consistent, predictive, and specific to the questions at issue. Genomics may not be useful in all situations or for all purposes in ERA or may not represent an improvement over existing methods in some areas. A variety of research needs and questions exists, including

- standardization of methods for collecting samples for exposure analyses to ensure sample integrity and lack of contamination
- development of basic exposure and genomic relational data or databases
- retrospective experiments that evaluate evidence for genomic-specific exposure to help confirm relationships between "fingerprints" and specific chemicals or classes of chemicals
- experiments to determine the relationship between actual exposure concentrations, tissue residue levels, and pattern of genomic responses to identify dose–response relationships and the potential of thresholds for genomic responses
- evaluation of the biological persistence of the genomic "signal" (i.e., how long the signal persists after an acute exposure or whether it changes over chronic exposures)
- experiments evaluating how food resource affects signal (e.g., if the compound is consumed through secondary exposure from prey items)
- whether genomics can provide indications of exposure when a compound does not accumulate
- the time course of genomic effects; how time affects measurement of signal
- determination of a signature of exposure in a mixture; ability to separate the signature of a single chemical within the broader signature of a mixture
- determination of the impact of environmental and organism conditions on genomic measurements; how water quality, temperature, season (daylight), and reproductive state affect the measurement (e.g., EROD and ECOD activities in relation to PAH exposure depending on the season)

4.3.1.2 Application of Genomics to Effects Characterization

Effect characterization describes the types of effects a pesticide can produce in an organism and how those effects change with varying pesticide exposure levels. This characterization is based on the available effects (toxicity) information for various plants and animals, including consideration of incident information and biological monitoring data. The toxicity testing scheme is tiered, with results from lower tiers used to determine potential harmful effects to nontarget organisms and the need for additional testing at higher tiers. Testing can progress from basic laboratory tests at the lowest level to applied field tests at the highest level. Table 4.3 lists the typical toxicity endpoints used as inputs to the risk quotient (RQ) method for calculating risk (USEPA 2004). The risk assessment process used by OPP (USEPA 2004) follows the broader USEPA ecological risk assessment guidelines and framework (USEPA 1998).

While the toxicity endpoints in Table 4.3 are routinely used to calculate RQs, they do not represent the entire universe of endpoints that may be considered in an ERA. The risk assessment may consider other available data that provide additional

TABLE 4.3
Typical Toxicity Testing Endpoints for Aquatic and Terrestrial Species Used in Pesticide Risk Assessments[a]

Aquatic organisms	Acute assessment	Lowest tested EC50 or LC50 for freshwater fish and invertebrates and estuarine or marine fish and invertebrates from acute toxicity tests
	Chronic assessment	Lowest NOEC for freshwater fish and invertebrates and estuarine or marine fish and invertebrates from early life stage or full life-cycle tests
Terrestrial organisms	Acute avian assessment	Lowest LD50 (single oral dose) and LC50 (subacute dietary)
	Chronic avian assessment	Lowest NOEC for 21-week avian reproduction test
	Acute mammalian assessment	Lowest LD50 from single oral dose test
	Chronic mammalian assessment	Lowest NOEC for 2-generation reproduction test
Plants	Terrestrial nonendangered	Lowest EC25 values from both seedling emergence and vegetative vigor for both monocots and dicots
	Aquatic vascular and algae	Lowest EC50 for both vascular and algae
	Terrestrial endangered	Lowest EC5 or NOEC for both seedling emergence and vegetative vigor for both monocots and dicots

[a] After Touart LW, Maciorowski AF. 1997. Ecol Appl 7(4):1086–1093.

information on existing toxicity endpoints, insight on endpoints not routinely considered for RQ calculation, and effects data on specific additional taxonomic groups (e.g., amphibian and freshwater mussel tests) not routinely tested during the ERA process. Professional judgment is used to determine whether or how data on other toxicological endpoints are included in the risk assessment.

Genomics may be best utilized in two important components of the effects characterization phase of the risk assessment process: defining mode of action and improving exposure–response relationships (e.g., identifying antecedents to significant apical effects at lower exposure concentrations). Numerous studies have demonstrated the ability of gene expression fingerprinting to discriminate between compounds with different pharmacological mechanisms or toxicological properties (Ellinger-Ziegelbauer et al. 2004; Kier et al. 2004). The development of large databases of chemical effects on gene expression has allowed the use of increasingly sophisticated data mining methods, including the use of classification algorithms such as support vector machines (SVMs) to derive sensitive and specific gene expression-based classifiers of biological response (Ganter et al. 2005; Natsoulis et al. 2005; Ruepp et al. 2005).

Such expression signatures can be earlier and more sensitive predictors of toxicity than existing clinical biomarkers (Fielden, Eynon, et al. 2005; Fielden, Pearson, et al. 2005) and can be derived for many diverse toxicological, pathological or pharmacological endpoints (Ganter et al. 2005). Further, the availability of expression profiles for chemicals of diverse biological properties allows the development of curated pathways of biological and toxicological response with gene expression data for the key protein components of those pathways. This facilitates the association of patterns of pathway response to chemical exposure with well-defined adverse effects related to that pathway. From the reverse perspective, interrogation of chemically induced gene perturbations with well-developed ontological annotations for the perturbed genes can expose novel or unanticipated mechanisms or modes of action.

The prospective approach to evaluation of the molecular status of cells and tissues requires a systematic course that begins by defining a toxicological pathway associated with a particular risk issue (Villeneuve et al. 2007). Once a pathway is identified as important, critical tissues and their respective roles in the pathway must be understood. This understanding is crucial to developing information on which functions are vulnerable or sensitive to toxicological perturbation. Once this understanding is developed, experiments are conducted with chemicals potentially modulating the vulnerable functionality. Examination of the transcriptomic, proteomic, and metabolomic responses from these experiments results in a molecular profile that contains the "signature" of the particular mechanisms of action. As the set of test chemicals is expanded, the molecular profile for a particular mechanism will increase in resolution, thus enabling a more focused assessment of the biological response.

Developing a computational model for predicting outcomes based on mechanism is an iterative process between model building and chemical testing. Initial models will aid in subsequent chemical selection and continued testing will improve these models. The key element to this approach is selection of chemicals for testing based on suspected mechanisms of action. The goal of this approach is to develop a computer-based computational model for predicting toxicity based on an understanding of how specific mechanisms of action affect the organism.

The improvement of exposure–response analysis (e.g., improving sensitivity, more conclusively linking specific chemical exposures with specific adverse outcomes) is another possible application for genomic technologies in risk assessment. Evaluation of exposure–response data is necessary at several points in the planning, formulation, and characterization phases of ERAs, including evaluating the exposure–response relationship, determining exposure sensitivity, establishing causality with ecological effects, and reducing the uncertainty of exposure-related extrapolations.

The current approach in pesticide registration for exposure–response data is to quantify endpoints (LC50, EC50, NOEC) for lethality and other established markers of toxicity at the organismal level, including growth or life-stage effects (reproductive or developmental abnormalities). While the final output of exposure–response assessments under the current paradigm is well established for integration into risk analysis equations, many of the tests measuring pathological and whole-animal toxicity response do not evaluate more subtle, lower dose toxicological effects and are likely to result in data gaps related to exposure.

Application of genomic technologies may provide an alternative approach to developing exposure–response relationships (including determination of relevant biomarkers and mechanism of action), reduce uncertainty of risk estimation associated with extrapolation to doses lower than those required to cause frank toxicity during acute or chronic tests (including measurement of more sensitive NOAELs), and, lastly, possibly establish a basis for predictive markers in short-term assays to replace the need for some chronic studies. These types of data have the potential to make exposure–response assessment simpler and more sensitive and relevant, thus providing some incentive for incorporation of the technology into the pesticide registration process.

4.3.1.3 Application of Genomics to Risk Characterization

Risk characterization is the integration of the effects and exposure characterizations to determine the ecological risk from the use of the pesticide and the likelihood of effects on aquatic life, wildlife, and plants based on varying pesticide use scenarios. Several uncertainties, which could be addressed through the use of genomic data, are frequently inherent in the ecological risk characterization of pesticides.

Lifestage sensitivity. It is generally recognized that test organism age may have a significant impact on the observed sensitivity to a toxicant. Fish acute toxicity data for ERAs are typically collected using juvenile fish between 0.1 and 5 g. Aquatic invertebrate acute testing is performed on recommended immature age classes (e.g., first instar for daphnids; second instar for amphipods, stoneflies, and mayflies; and third instar for midges). Similarly, acute dietary testing with birds is also performed on juveniles, with mallards being 5 to 10 days old and quail 10 to 14 days old.

Testing of juveniles may overestimate toxicity at older age classes for pesticide active ingredients that act directly (i.e., without metabolic transformation) because younger age classes may not have developed the enzymatic systems necessary to detoxify xenobiotic compounds. However, the influence of age may not be uniform for all compounds, and compounds requiring metabolic activation may be more toxic in older age classes. The risk assessment process has no current provisions for a methodology that would account for this uncertainty. Insofar as the available toxicity data may provide ranges of sensitivity information with respect to age class, the risk assessment uses the most sensitive life-stage information as the conservative endpoint and includes an evaluation of all available age-class sensitivity information as it affects the confidence of risk conclusions during risk characterization.

The massively parallel analysis of gene products, protein products, and metabolites via genomic technologies holds great promise for addressing the issue of age-class sensitivity. Since divergent chemical classes may elicit maximal response at different life stages in target organisms (depending on the mechanism of action), it should be possible to use genomic approaches to define differences in response across age classes within a particular test species since genomic technologies directly interrogate the molecular basis of biological response to a chemical exposure. Two principal questions can be addressed by such experiments:

- Genomic and/or metabolomic fingerprinting of response at a given chemical dose across species life stages (from embryo to adult) can identify whether

molecular-level changes (perhaps tied to pathways, in cases of well-annotated genomes) are conserved temporally through the life cycle of the test organism. If not conserved, they can help identify when the organisms begin to become (or stop being) responsive at the biomolecular level.
- Quantitative analysis of the changes in gene expression or metabolite flux at the different life stages can provide information on the most sensitive biomarker choice for a given life stage and a given test chemical.

Overall, these experiments would allow risk assessors to design studies to reduce uncertainty related to the differential sensitivity of age class within the context of a single species.

The use of genomics to define life-stage sensitivities will likely involve pilot-scale, "discovery" types of studies. For a given species it should only be necessary to perform transcript, protein, or metabolite fingerprinting experiments at the beginning of a particular chemical's evaluation cycle. Once these studies have been done, further definitive experimentation and testing can be performed using the chosen life stage and with appropriate biomarkers or endpoints, guided mechanistically by the pilot genomic data. This paradigm represents a particularly powerful and efficient use of data-rich, resource-intensive, and exploratory transcriptomic, proteomic, and metabolomic technologies. In addition, it will be important to use knowledge of putative mode of action for particular chemicals in combination with genomic approaches to drive choice of test organism and organism age class. In cases where the mechanism of action is suspected for a particular pesticide, it may be possible to use previously collected genomic or metabolomic data for compounds of similar structure or compound class to inform choice of test species life stage, rather than conducting full, genome-enabled pilot screening.

Particular research considerations for the use of genomics in defining species age-class sensitivity include availability of genomic tools for the test species, annotation and classification of genes within specific pathways, and choice of appropriate test tissues and exposure regimes for a given pesticide chemical. In general, efforts should be made to reduce the number of species to be included in genome-enabled, age-class testing for pesticide risk assessment. Sequencing projects currently exist for some aquatic species (e.g., fathead minnow and daphnia) and it is suggested that these species may be good choices for initial focus on transcriptomic (e.g., microarray-based) characterization of compound sensitivity by age class. For terrestrial species, *Caenorhabditis elegans* is an obvious candidate as details on the transcription status of worms in the different larval and adult stages are already known (Jiang et al. 2001). These considerations are most important for genomic and proteomic fingerprinting, as such experiments depend explicitly on availability of sequence data. Metabolomic fingerprinting is much less dependent on sequence information. Thus, given sufficient sensitivity, it should be possible to conduct age-class-specific profiling of contaminant response in nonmodel organisms (e.g., quail, sheepshead minnow) using metabolomic methods.

An important consideration for proteomic and metabolomic fingerprinting by age class is sample size limitation. For these experiments, analyses would need to be conducted over a wide range of organism size and it might be difficult in many cases to maintain organ specificity from embryo to adult when isolating tissues.

Sensitivity becomes more of an issue for proteomic and metabolomic analyses than for transcriptional analysis because, for the latter, amplification of analytes is possible. Technical challenges related to sample preparation (e.g., protein extraction and separation techniques) and analysis (e.g., mass spectrometric and NMR sensitivity) need to be addressed in order to apply these techniques readily to small samples (e.g., embryos) in a high-throughput manner. The development of technical solutions such as microprobe technology for NMR, however, suggests that issues concerning sample size can be dealt with satisfactorily.

Species sensitivity. Experiments designed to map sensitivities within species (along the temporal life-stage axis) using genomic technologies would be valuable, in combination with cross-species sensitivity comparisons, for increasing the understanding of potential ecosystem-wide effects of pesticides. The uncertainty inherent in choosing a specific model organism as the basis of an ecological risk assessment should decrease as transcriptomic, proteomic, and metabolomic fingerprints are developed for specific compound classes across species and life stages of interest.

The ERA process for conventional outdoor-use pesticides requires toxicity data for a range of aquatic and terrestrial vertebrates and invertebrates. The ERA then relies on a toxicity endpoint from the most sensitive species tested as the basis for regulation. However, the endpoint for the most sensitive test species tested does not necessarily reflect the sensitivity of the most sensitive species in a given environment. The sensitivity ranking of the most sensitive tested species in relation to the distribution of sensitivities for all possible species is a function of the number of species tested, as well as the variability among all species to a particular chemical. There is currently considerable uncertainty regarding the sensitivity of wild species, particularly threatened and endangered species, in comparison to standard laboratory test species such as those in Table 4.1. It has been argued that species that can be easily cultured and maintained in the laboratory could have a lower inherent sensitivity to toxicants and other stressors when compared to wild species that have specific habitat requirements, and are thus difficult or impossible to keep under laboratory conditions. Whether this is the case or not, it is certainly true that the amount of knowledge concerning the inherent variability of sensitivity among all species versus traditional toxicity test species is limited.

One approach to dealing with uncertainty in comparative species sensitivities between tested and wild species is through the use of safety factors. However, recognition is growing that the use of safety factors can result in conservative or erroneous risk estimates and consequently waste resources. The issue of reliable extrapolation of species sensitivity from test species to untested species has proved one of the most difficult issues to resolve in regulation. Ecotoxicogenomics potentially offers a number of approaches and methods that could support better risk assessment in this regard. One of the immediate benefits of genomic approaches is that nondestructive samples for wildlife species of concern can be used in many cases to develop information on organism exposure if relevant genomic patterns for a particular chemical or class of chemicals can be identified in closely related surrogate organisms and confirmed in the species of concern.

In the short term, it may be feasible to conduct a limited set of cross-species experiments to collect comparative responses to a given chemical supported by existing toxicological data. This kind of approach could utilize the concept of a single discriminating dose and focus on changes in expression of signature gene sets whose expression change is known to be subject to perturbation by chemicals of different classes and modes of action. Linking the strength and direction of change in signature genes to the strength of the toxic effect at a discriminating dose will provide information regarding the species (or broader phyla) that are inherently most sensitive to the particular category of chemical. This would allow regulators to target their requests for data, with the result that fewer tests may be needed and the data gathered would better fit the needs of the ERA.

The next step in the interspecies extrapolation would be data collection from individuals exposed to a carefully selected set of pesticides. The chemicals should be chosen to represent the dominant and most problematic compounds used in the agriculture, preferably those for which a wealth of information collected by the traditional methods is available. Following the selection of optimal pesticide dose and developmental stage, low-dose chronic exposure studies should be conducted and samples collected throughout the experiments for evaluation of genomic changes. Response patterns resulting from these experiments should be anchored to resulting phenotypes (endpoints) such as reduced growth rate, decreased reproductive fitness, etc. Since mechanistic comparisons may be limited to particular pathway genes, it may be better to use more targeted expression analysis systems (targeted arrays, Q-PCR) to determine expression of signature genes.

In addition to the cautious use of cross-species large-scale or targeted transcriptomic and proteomic approaches, comprehensive metabolic fingerprinting coupled with multivariate data handling can offer a complementary approach to transcriptional and protein fingerprinting in the detection of change in sensitive pathways. Metabolomics offers enormous potential for the comparative biochemical assessment of related and unrelated nonmodel organisms. Metabolomics does not have the limitation of needing species-specific genome elucidation. There also are few analytical restrictions on measuring low molecular weight metabolites from any species, and the common nature of some primary metabolites allows direct comparisons of biochemical changes. Mass spectrometry-based metabolomics and 1H NMR can provide data that can be used to identify metabolic perturbations. Multivariate data handling methods can be used to detect change in metabolite profiles following chemical exposure. Exploiting discriminate analysis models across species can also enable comparison of putative biomarker responses detected in one species to the responses in a second species to determine if the pathway is more or less perturbed.

4.4 RISK ASSESSMENT FOR ANTIMICROBIALS AND INERT INGREDIENTS

4.4.1 Antimicrobials

4.4.1.1 Current Practices

The USEPA practice for assessing risk for antimicrobials is similar to the assessment of pesticide active ingredients (Section 4.3, this chapter), which uses a tiered system of ecological effects testing to assess the potential risks to nontarget plants, aquatic and terrestrial vertebrates and invertebrates, and nontarget insects. These tests advance from basic acute, subacute, and reproductive (chronic) laboratory tests to field tests. The results of each set of tests must be evaluated to determine the potential of the antimicrobial compound to cause adverse effects and the potential need for any additional testing.

Typically, only a small set of ecological effects and environmental fate data are requested, given the low prospect of significant environmental exposure for many antimicrobial uses. These studies characterize hazard to target species for label hazard statements and in case of an unexpected release (spill) to the environment. Testing of additional species or higher tier testing can be requested based on the results of this basic set of studies or reports of adverse effects in the literature. Under this approach, registrants of low-exposure antimicrobials may perform tests in a tiered fashion.

Antimicrobials used for aquatic areas or industrial processes and water systems, antifoulants, and wood preservatives require data sets similar to those described for conventional pesticides. These uses occur outdoors, discharge effluent directly to the outdoors, or result in materials treated with antimicrobials (i.e., wood preservatives and antifoulants) being placed in the environment, thereby leading to potentially significant environmental exposure.

4.4.2 Inert Ingredients

4.4.2.1 Current Practices

The USEPA has developed procedures for assessing toxicity and risks associated with inert ingredients (chemicals in the pesticide formulation that are not the principal active pesticide—i.e., solvents) and surfactants in pesticide formulations (USEPA 2004) and performs either qualitative or quantitative assessments of potential risk associated with these chemicals. The decision to perform either type of assessment is based on available information on the chemical and toxicological characteristics of the inert ingredient. Information on toxicological characteristics may include available ecotoxicology data from the literature, information on closely related chemical analogues, and/or effects estimated from quantitative structure–activity relationships (QSAR). Use of structural analogy or QSAR models is consistent with techniques employed in other USEPA programs (e.g., Office of Pollution Prevention and Toxics [OPPT] premanufacturing notification [PMN] process). This information is used to determine if inert ingredients can be classified as 1) generally recognized as safe, 2) available data insufficient to confirm little or no toxicity and additional study required, or 3) sufficient toxicological and exposure concern to warrant a quantitative risk assessment similar to those conducted for pesticide active ingredients.

A tiered methodology is used for evaluating low- or low-to-moderate toxicity chemical substances. This tiered methodology facilitates decisions, in a streamlined manner, for low- or low-to-moderate-toxicity chemical substances in order to focus resources on those chemicals of potentially higher toxicity requiring in-depth evaluation.

Tier 1 chemicals are presumed to have low- or low-to-moderate toxicity for which there is readily available scientifically valid information or data to make a confirmatory judgment of this presumption. Tier 2 chemicals do not have sufficient information to assess the chemical substance's toxicity. Given the lack of readily available information to classify a chemical as Tier 1, OPP requires the submission of a limited data set to characterize the hazard of the chemical substance. This data set is similar to the internationally recognized OECD screening information data set (SIDS). Chemical substances that appear to have appreciable toxicity would be assigned to Tier 3, requiring a complete (food-use) 40 CFR Part 158 database to make these safety determinations.

4.4.3 Application of Genomics to ERA for Inert Ingredients and Antimicrobials

Much of the discussion in Section 4.3 of this chapter is directly applicable to ERA for antimicrobials and inert ingredients. One key area of need where the application of genomic technology can immediately begin to furnish practical solutions is in the identification of possible toxicities associated with compounds for which little or no safety data currently exist, such as inert ingredients. The use of toxicogenomics to predict, identify, and provide mechanistic information on likely toxic responses to key chemical classes will allow the adoption of an intelligent, scientific, data-driven approach to prioritization of compounds for which additional toxicity testing may be required.

Genomic techniques could allow for the introduction of targeted in vitro test systems into the practice of toxicology. One of the promising possibilities is the analysis of cell and/or tissue culture for the evaluation of many different possible mechanisms of toxicity with a single molecular profile. Initial experiments of this kind could be performed on a global scale; however, at later stages smaller, targeted mechanistic patterns could be used to analyze certain cells or tissues. These targeted analyses would contain a subset of molecular probes highly relevant to the metabolic pathways essential to the function of the cell or tissue type. The advantage of such analyses would be that treatments could be performed within a relatively short time and the researcher would not need to sift through thousands of molecular endpoints when analyzing the results produced. Once the results of such quick in vitro tests were added together, their summary would be expected to provide a reasonable estimate for the outcome of the in vivo test. The biggest potential disadvantage of such an approach would be the loss of effects from metabolites produced in other tissues or organs. However, in the absence of any data, such as in the case for inert ingredients, the resulting in vitro data would be of considerable value.

4.5 NECESSARY DEVELOPMENTS IN APPLICATION OF GENOMIC APPROACHES TO RISK ASSESSMENT OF PESTICIDES, ANTIMICROBIALS, AND INERT INGREDIENTS

4.5.1 Establish Baseline Variation of the Transcriptome, Proteome, and Metabolome

The most basic need is to calibrate the genomic technologies in a given species by setting up historical control baselines to determine the variation of expression profiles for the various genomic measurements. This is especially important for those species for which near-isogenic strains are not available in the laboratories, as published data show that the range for certain mRNA (and, presumably, protein) levels could be highly variable in individuals of natural populations (see, for example, Oleksiak et al. 2002, 2005). This variation will also be present even in inbred, near-isogenic populations kept in the lab for generations, albeit at a much lower level. Once the range of variation in "normal" laboratory control populations is determined, pesticide effects at the level of the transcriptome will be more clearly identifiable from background variation. It will also be important, however, to evaluate the variability in laboratory populations under controlled conditions versus the variability of natural populations that reflect the entire range of environmental factors that could affect variability in responses.

It is expected that the gene (protein) content of microarrays for different species would overlap to a large extent, provided that the source of the sample collection and processing were standardized. These overlapping gene sets would allow direct comparisons of the effects of the same pesticide in all the species tested and would allow the researcher to rank species in the order of their relative sensitivity. The presence or absence of specific pathways can be determined to facilitate evaluation of comparative responses to pesticides. This treatment series is expected to determine the "genomic sensitivity" levels of the organisms to particular classes of pesticides. These responses will not necessarily be the same as those obtained with the traditional tools. The subsequent results would allow researchers to determine the sensitivity and predictive ability of the genomic technologies in comparison to pre-existing toxicological tools.

4.5.2 Cross-Species Experiments Using Existing Tools

A major challenge is that there is only a limited set of species for which sufficient genomic sequence information is available to conduct thorough transcriptomic and proteomic analysis. This means that transcriptomic and proteomic platforms specific for all the ecologically relevant model species cannot be developed at this time. Thus, in the short term, genomic research aimed at understanding cross-species extrapolation and species sensitivity issues will be limited by the coverage of existing tools. During the initial phase of the genomic era, DNA (and protein) chips are going to be available only for a limited number of "sequence-rich" species for which a large number of full-length cDNA or expressed sequence tag (EST) sequences have been described. For teleost fishes, which are in fact one of the taxa for which there is a "comparative" wealth of sequence data available, the species currently covered include only the rainbow trout, roach, common carp, zebrafish, fathead minnow, and medaka. For other taxa, the situation is often much worse (e.g., for amphibians, sufficient sequence data are only available for two tropical frog species: *Xenopus laevis* and *X. tropicalis*).

At this stage, cross-species array hybridization (i.e., applying labeled targets from a species onto an array originally developed for another species) also could be used to extend the testing ability to those species for which sequence information and consequently arrays are not currently available. Published data from the peer-reviewed literature indicate that such cross-species hybridization onto microarrays is feasible between not only closely related species (e.g., Locke et al. 2003; Gilad et al. 2005) but also occasionally distantly related ones as well (e.g., Renn et al. 2004).

The extent to which cross-species array hybridization can be applied to the analysis of distantly related species must be determined experimentally. Although the increasing evolutionary distance results in a decreasing number of responding genes on the array, the remaining ones would still allow for some degree of quick testing until the tools for the "sequence-poor" species are developed. The use of the same array on several species would allow for extrapolation of results across the taxonomic range, provided that the technical challenges associated with cross-species hybridization can be resolved. The extent to which cross-species array hybridization can be applied to the analysis of distantly related species, however, must be determined experimentally in taxa for which phylogenetic relationships have been robustly investigated and therefore evolutionary relationships between species are known.

4.5.3 Development of Genomic Tools for the Most Ecologically Relevant Model Species

From a longer term perspective, although cross-species hybridization onto cDNA microarrays can help to extend coverage to "sequence-poor" species, tools are most critical for the important species for risk assessment purposes. Currently, the fastest and most affordable means to collect such sequence information is to generate ESTs. These sequences are produced by end-sequencing randomly pulled cDNA clones from libraries; clustered EST sets containing sequences from a few thousand clones allow for quick generation of arrays from "sequence-poor" organisms without sequenced genomes (Chen et al. 2004). Genome projects aiming at low level (0.1 to 2×) coverage are expected to fill the gaps and yield sequence information necessary for developing arrays in additional "toxicogenomic" species. Ideally, the selection of the species for such analysis should consider their taxonomic position as well as their potential usefulness in the risk assessment process and the size of their genomes.

Once data generated by species-specific tools (e.g., microarrays or protein arrays) from several species are available, comparative analysis becomes possible. Expression profiles can be analyzed and compared and pathway-specific responses can be identified. For the identification of pathways, there is a pressing research need for the development of a common gene ontology to simplify the read-across of transcriptional responses between closely and more distantly related species. Notwithstanding this issue, the production of such extended data sets would create the theoretical possibility of predicting the effects of a given pesticide on a "sequenceless" species by interspecies extrapolation. Transcription profiles from several different species treated with a chemical could be analyzed to identify key pathways and genes with the highest response intensity to treatment. Conserved regions could be selected and used to design degenerate primers for the amplification and isolation of the orthologs from the new species. These orthologs could then be spotted onto a slide to form a targeted mini-array and analyzed on the new species and on several characterized species treated with the chemical. Alternatively, comparison could also be made by focused Q-PCR analysis.

4.5.4 Need for Large-Scale Demonstration Projects

To appreciate and investigate the potential for cross-species comparisons and particularly to improve the scientific basis for the extrapolation of data from standard test species to untested organisms, a set of suitable demonstration projects needs to be supported. There are many cases in the literature where clear sensitivity differences between species have been detected (e.g., Spurgeon et al. 2000). Understanding the mechanistic basis of such differences in sensitivity for pesticides would require a coordinated 3-stage approach that includes: assessment of the influence of biogeochemical and physiological interactions on exposure, uptake, and internal dose, assuming that sensitivity differences cannot be accounted for by differences in exposure; evaluation of the consequences of first- and second-phase metabolism for internal exposure; and quantification and comparison of signature changes linked to the observed toxicological responses.

Genomics has the potential to bring radical change to the generation of data and preparation of pesticide dossiers for risk assessment purposes. From a regulatory perspective, the full realization of this potential is likely to be 5 to 10 years in the future. However, as previously indicated, one key area where the application of genomic technology can immediately begin to furnish practical solutions is in the identification of possible toxicities associated with compounds for which little or no safety data currently exist, such as inert ingredients and antimicrobials. The use of toxicogenomics to predict, identify, and provide mechanistic information on likely toxic responses for these chemicals would bolster the current data generation for many of these compounds. It would also provide a proving ground for developing the confidence in the techniques necessary to transfer the approaches to pesticides where the data generation requirements and associated costs are greater.

REFERENCES

Barton HA, Pastoor TP, Baetcke K, Chambers JE, Diliberto J, Doerrer NG, Driver JH, Hastings CE, Iyengar S, Krieger R, et al. 2006. The acquisition and application of absorption, distribution, metabolism, and excretion (ADME) data in agricultural chemical safety assessments. Crit Rev Toxicol 36:9–35.

Bradbury SP, Feijtel TCJ, Van Leeuwen CJ. 2004. Meeting the scientific needs of ecological risk assessment in a regulatory context. Environ Sci Technol 38(23):463A–470A.

Bulera SJ, Eddy SM, Ferguson E, Jatkoe TA, Reindel JF, Bleavins MR, De La Iglesia FA. 2001. RNA expression in the early characterization of hepatotoxicants in Wistar rats by high-density DNA microarrays. Hepatology 33(5):1239–1258.

Carmichael NG, Barton HA, Boobis AR, Cooper RL, Dellarco VL, Doerrer NG, Fenner-Crisp PA, Doe JE, Lamb JC, Pastoor TP. 2006. Agricultural chemical safety assessment: a multisector approach to the modernization of human safety requirements. Crit Rev Toxicol 36:1–7.

Chen YA, McKillen DJ, Wu SY, Jenny MJ, Chapman R, Gross PS, Warr GW, Almeida JS. 2004. Optimal cDNA microarray design using expressed sequence tags for organisms with limited genomic information. BMC Bioinformatics 5:191.

Cooper, RL, Lamb JC, Barlow SM, Bentley K, Brady AM, Doerrer NG, Eisenbrandt DL, Fenner-Crisp PA, Hines RN, Irvine LFH, et al. 2006. A tiered approach to life stages testing for agricultural chemical safety assessment. Crit Rev Toxicol 36:69–98.

Denslow ND, Kocerha J, Sepulveda MS, Gross T, Holm SE. 2004. Gene expression fingerprints of largemouth bass (*Micropterus salmoides*) exposed to pulp and paper mill effluents. Mutat Res 552(1–2):19–34.

Doe JE, Boobis AR, Blacker A, Dellarco VL, Doerrer NG, Franklin C, Goodman JI, Kronenberg JM, Lewis R, McConnell EE, et al. 2006. A tiered approach to systemic toxicity testing for agricultural chemical safety assessment. Crit Rev Toxicol 36:37–68.

Ellinger-Ziegelbauer H, Stuart B, Wahle B, Bomann W, Ahr HJ. 2004. Characteristic expression profiles induced by genotoxic carcinogens in rat liver. Toxicol Sci 77:19–34.

European Commission. 2005. REACH and the need for intelligent testing strategies. Joint Research Centre, Institute for Health and Consumer Protection, 26 p.

Fielden MR, Eynon BP, Natsoulis G, Jarnagin K, Banas D, Kolaja KL. 2005. A gene expression signature that predicts the future onset of drug-induced renal tubular toxicity. Toxicol Pathol 33(6):675–683.

Fielden MR, Pearson C, Brennan R, Kolaja KL. 2005. Preclinical drug safety analysis by chemogenomic profiling in the liver. Am J Pharmacogenom 5(3):161–171.

Forbes V, Palmquist A. 2006. The use and misuse of biomarkers in ecotoxicology. Environ Toxicol Chem 25:272–280.

Ganter B, Tugendreich S, Pearson CI, Ayanoglu E, Baumhueter S, Bostian KA, Brady L, Browne LJ, Calvin JT, Day G-J, et al. 2005. Development of a large-scale chemogenomics database to improve drug candidate selection and to understand mechanisms of chemical toxicity and action. J Biotechnol 119(3):219–244.

Gilad, Y, Rifkin, SA, Bertone, P, Gerstein, M, White, KP. 2005. Multispecies microarrays reveal the effect of sequence divergence on gene expression profiles. Genome Res 15:674–680.

Hamadeh H, Bushel PR, Jayadev S, Martin K, Sorbo O, Sreber S, Bennet L, Tennant R, Stoll R, Barrett JC, et al. 2002. Prediction of compound signature using high-density gene expression profiling. Toxicol Sci 67:232–240.

Jiang M, Ryu J, Kiraly M, Duke K, Reinke V, Kim, SK. 2001. Genome-wide analysis of developmental and sex-regulated gene expression profiles in *Caenorhaditis elegans*. Proc Natl Acad Sci USA 98:218–223.

Kier LD, Neft R, Tang L, Suizu R, Cook T, Onsurez K, Tiegler K, Sakai Y, Ortiz M, Nolan T, et al. 2004. Applications of microarrays with toxicologically relevant genes (tox genes) for the evaluation of chemical toxicants in Sprague Dawley rats in vivo and human hepatocytes in vitro. Mut Res 549:101–113.

Klaper R, Thomas M. 2004. At the crossroads of genomics and ecology: the promise of a canary on a chip. BioScience 54:403–412.

Larkin P, Sabo-Attwood T, Kelso J, Denslow ND. 2003. Analysis of gene expression profiles in largemouth bass exposed to .17-beta-estradiol and to anthropogenic contaminants that behave as estrogens. Ecotoxicology 12(6):463–468.

Lindon JC, Nicholson JK, Holmes E, Antti H, Bollard ME, Keun H, Beckonert O, Ebbels TM, Reilly MD, Robertson D, et al. 2003. Contemporary issues in toxicology—the role of metabonomics in toxicology and its evaluation by the COMET project. Toxicol Appl Pharmacol 187:137–146.

Locke DP, Segraves R, Carbone L, et al. 2003. Large-scale variation among human and great ape genomes determined by array comparative genomic hybridization. Genome Res 13:347–357.

Natsoulis G, El Ghaoui L, Lanckriet GR, Tolley AM, Leroy F, Dunlea S, Eynon BP, Pearson CI, Tugendreich S, Jarnagin K. 2005. Classification of a large microarray data set: algorithm comparison and analysis of drug signatures. Genome Res 15(5):724–736.

Oleksiak MF, Churchill GA, Crawford DL. 2002. Variation in gene expression within and among natural populations. Nat Genet 32:261–266.

Oleksiak MF, Roach JL, Crawford DL. 2005. Natural variation in cardiac metabolism and gene expression in *Fundulus heteroclitus*. Nat Genet 37:67–72.

Renn SCP, Aubin-Horth N, Hoffmann H. 2004. Biologically meaningful expression profiling across species using heterologous hybridization to a cDNA microarray. BMC Genomics 5:42.

Royal Commission on Environmental Pollution. 2003. Chemicals in products: safeguarding the environment and human health, CM 5827. London: Royal Commission on Environmental Pollution.

Ruepp S, Boess F, Suter L, de Vera MC, Steiner G, Steele T, Weiser T, Albertini S. 2005. Assessment of hepatotoxic liabilities by transcript profiling. Toxicol Appl Pharmacol 207(2 Supp l):S161–S170.

Sawada H, Takami K, Asahi S. 2005. A toxicogenomic approach to drug-induced phospholipidosis: analysis of its induction mechanism and establishment of a novel in vitro screening system. Toxicol Sci 83(2):282–292.

Spurgeon DJ, Svendsen C, Rimmer VR, Hopkin SP, Weeks JM. 2000. Relative sensitivity of life-cycle and biomarker responses in four earthworm species exposed to zinc. Environ Toxicol Chem 19:1800–1808.

Steiner G, Suter L, Boess F, Gasser R, de Vera MC, Albertini S, Ruepp S. 2004. Discriminating different classes of toxicants by transcript profiling. Environ Health Perspect 112(12):1236–1248.

Touart LW, Maciorowski AF. 1997. Information needs for pesticide registration in the United States. Ecol Appl 7(4):1086–1093.

Tugendreich S, Pearson CI, Sagartz J, Jarnagin K, Kolaja K. 2006. NSAID-induced acute phase response is due to increased intestinal permeability and characterized by early and consistent alterations in hepatic gene expression. Toxicol Pathol 34(2):168–179.

[USEPA] US Environmental Protection Agency. 1998. Guidelines for ecological risk assessment. US Environmental Protection Agency, Risk Assessment Forum, Washington, DC, EPA/630/R095/002F.

[USEPA] US Environmental Protection Agency. 2000. The use of data on cholinesterase inhibition for risk assessments of organophosphorus and carbamate pesticides. Science Policy. Office of Pesticide Programs, EPA, Washington DC.

[USEPA] US Environmental Protection Agency. 2004. Overview of the ecological risk assessment process in the office of pesticide programs, US Environmental Protection Agency, endangered and threatened species effects determinations. Office of Prevention, Pesticides and Toxic Substances Office of Pesticide Programs, Washington, DC.

[USFDA] US Food and Drug Administration. 2005. Innovation/stagnation: challenge and opportunity on the critical path to new medical products. http://www.fda.gov/oc/initiatives/criticalpath/whitepaper.html

Villeneuve DL, Larkin P, Knoebl I, Miracle AL, Kahl MD, Jensen KM, Makynen EA, Durhan EJ, Carter BJ, Denslow ND, et al. 2007. A graphical systems model to facilitate hypothesis-driven ecotoxicogenomics research on the teleost brain–pituitary–gonadal axis. Environ Sci Technol 41:321–330.

Yang Y, Abel SJ, Ciurlionis R, Waring JF. 2006. Development of a toxicogenomics in vitro assay for the efficient characterization of compounds. Pharmacogenomics Mar 7(2):177–186.

5 Application of Genomics to Assessment of the Ecological Risk of Complex Mixtures

Edward J Perkins, Nancy Denslow, J Kevin Chipman, Patrick D Guiney, James R Oris, Helen Poynton, Pierre Yves Robidoux, Richard Scroggins, and Glen Van Der Kraak

CONTENTS

5.1	Background	88
5.2	Environmental Monitoring in a General Ecological Risk Assessment Framework	89
	5.2.1 Current Approaches to Monitoring Exposure and Effects	92
	5.2.2 Measurements of Biological Exposure and/or Effects	92
	5.2.3 Examples of Current Approaches to Exposure and Effects Monitoring	93
	5.2.4 Assessing Exposure and Effects with Mixtures of Contaminants	94
	5.2.5 Current Challenges in Environmental Monitoring and Assessment	96
5.3	The Promise of Genomics in Monitoring	97
	5.3.1 Potential Contributions of Genomics to Regulatory Assessment	98
5.4	Application of Genomics in Environmental Monitoring	99
	5.4.1 Using Genomics to Aid in Problem Formulation and Hypothesis Generation: Tier I	99
	5.4.1.1 Genomics Can Assist in Problem Formulation and Hypothesis Generation in Real-World Situations: Examples from Several Current Studies	99
	5.4.2 Genomics and Exposure and Effects Screening and Monitoring: Tier II	101
	5.4.2.1 Development of Screening Tools for Exposure and Effects Using Genomics	101

		5.4.2.2	Biological Significance: Linking Changes at the Molecular Level with Adverse Effects 102

	5.4.3	Assessing Acute and Chronic Effects with Genomics: Tier III 103
	5.4.4	Using Genomics in Assessments on Complex, Case-Specific Conditions: Tier IV 105

        5.4.4.1 Understanding Chemical Mixture Effects and Interactions of Multiple Stressors 105
        5.4.4.2 Defining Modes and Mechanisms of Action Can Answer Many Questions 106
        5.4.4.3 Determining Impacted Functions through Pathway Analysis 107
        5.4.4.4 Extrapolation of Chemical Effects across Many Species 107
        5.4.4.5 Reducing Uncertainty in Population Impacts by Defining Distribution of Individual Sensitivities 108

5.5 Developing Genomics for Regulatory Monitoring 109
5.6 The Outlook for Genomics and Environmental Monitoring 110
    5.6.1 Increasing Availability of Complete Genomes, Annotation, and Arrays 110
    5.6.2 Development and Maintenance of Databases Support Application of Genomics to Monitoring 111
    5.6.3 The Future of Genomics in Understanding Effects: Systems Toxicology 112
References 113

5.1 BACKGROUND

The environment has been increasingly impacted by chemical, biological, and physical stressors of anthropogenic origins that can ultimately result in degradation of habitat, loss of species, reduction of natural resources, and, potentially, adverse human health effects. Conditions where chemicals pose threats in the environment are complex and varied. Rarely occurring independently of one another, different chemical stressors are often found as mixtures in water, sediment, and soil. Animals are subjected to changing environmental and physiological conditions that may modulate chemical toxicity. Assessing toxicity is not only dose dependent but is also complicated by length of exposure and developmental stage. Environmental monitoring efforts, in both a regulatory and a status and trends framework, provide opportunities to assess stressors before they are manifested in populations and cause irreparable damage to the environment.

    This chapter seeks to explore ways by which genomics may add value to current approaches in environmental monitoring. In the context presented here, environmental monitoring refers to activities where field samples are taken, analyzed, and used for making regulatory decisions as in ecological risk assessments, for generating prioritization indices, and/or to develop status and trends information to better manage ecological health. We will first describe environmental monitoring in a general ecological risk assessment framework, describe current challenges in monitoring and assessment, and present potential uses of genomics (genetics, transcriptomics, proteomics, and metabolomics) in monitoring. Current limitations of genomics approaches are also discussed. The application of genomics in environmental monitoring will then be described in a generic tiered approach format commonly used in ecological risk assessments. Descriptions and definitions of specific terms and technologies are addressed in Chapter 1 and the book's Glossary.

Environmental monitoring efforts utilize a spectrum of approaches to target individuals, populations, communities, and, ultimately, ecosystems to detect and assess impacts. Monitoring of environmental effects is especially important in an adaptive risk or resource management framework (Figure 5.1). Environmental monitoring approaches are diverse and can include one or more of the following endpoints: quantification of key biochemical or biological effects; toxicity determination using bioassays with field-collected media or caged organisms; chemical abundance in tissues, soil, sediment, and/or water; reproductive success and growth; gene flow; migration patterns; changes in species abundance and diversity; and, on the largest scale, chemical and nutrient dynamics of ecosystems.

5.2 ENVIRONMENTAL MONITORING IN A GENERAL ECOLOGICAL RISK ASSESSMENT FRAMEWORK

Environmental monitoring is an integral part of assessing risks that exposure to chemicals and other stressors present to ecological receptors, including both vertebrates and invertebrates. Determining how much risk chemicals pose in the environment is a complex endeavor requiring problem definition, assessing potential and actual exposure and effects, and estimating the risk given to expected exposures. A multiphase or tiered risk assessment framework (Figure 5.1) provides a logical flow that utilizes monitoring to improve the ecological assessment process. Environmental monitoring and ecological risk assessments are generally conducted due to regulatory demands or to a need to monitor the conditions of specific resources or populations. Each of these has different characteristic needs that drive how monitoring is performed.

Monitoring and assessment in a regulatory context drive many monitoring programs. Several countries have environmental monitoring programs prescribed through various legislative orders, directives, or acts to set regulations or address environmental concerns (Table 5.1). Regulatory monitoring occurs in programs where industry, municipalities, or specific groups are mandated to ensure proper characterization and control of releases of contaminants or other stressors to the environment. Enforcing parties are required to protect resources and uphold regulatory standards. Typically, enforcing parties monitor responsible parties (i.e., manufacturing plants and others) to assure attainment of regulatory standards; consequences of fines, litigation, or corrective actions are levied against the responsible party for nonattainment.

Because of the often litigious nature of regulatory monitoring, methods and approaches in a regulatory context usually are highly prescribed and directed to

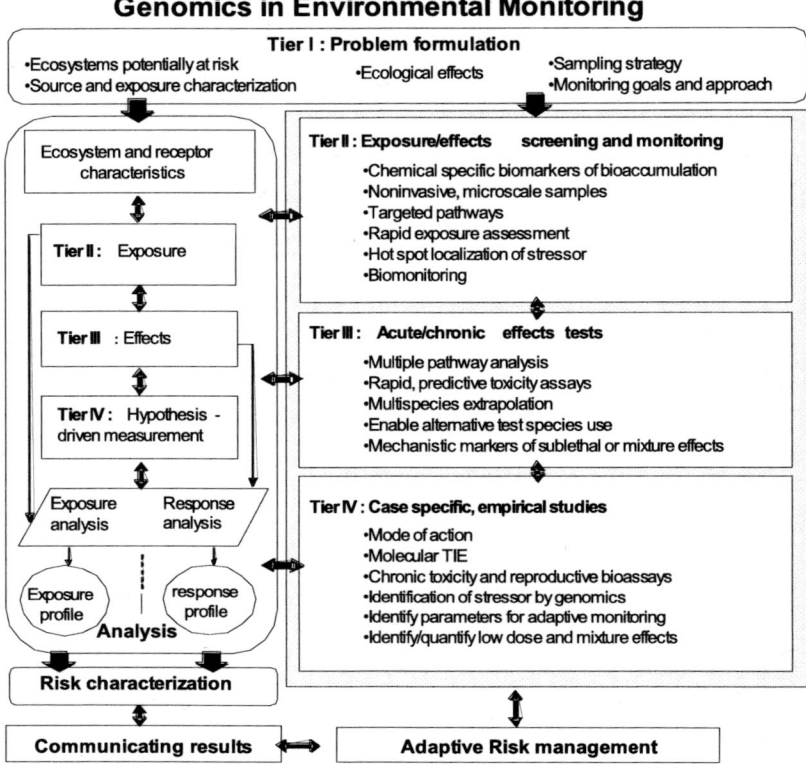

FIGURE 5.1 Integration of genomics with monitoring in a tiered assessment. The ecological risk assessment framework is shown as a tiered process. (Taken from Bradbury SP et al. 2004. Environ Sci Technol 38:463A–470A. Reproduced with permission from Environ Sci Technol. Copyright 2004 Am Chem Soc.)

assessing levels of specific chemicals. Regulatory agency personnel must ultimately understand and utilize monitoring data in order to establish permissible chemical or stressor levels. Sufficient guidance must be available to interpret monitoring data and determine how the data relate to adverse effects in the environment. From a regulatory point of view, unless endangered or threatened species are involved, a chemical must negatively impact population survival to pose an environmental risk. Therefore, growth, reproduction, and survival are often endpoints in assessing risk once exposure is detected through monitoring efforts.

Status and trend monitoring is an assessment protocol where environmental samples are collected, typically by government agencies, to track spatial or temporal trends of contaminants and other stressors or to understand the dynamics of natural fauna and flora populations or communities. Status and trend monitoring programs establish baseline or background conditions and levels of contaminants and other stressors in water, sediment, soil, air, or biota. Regional, national, and international contaminant monitoring programs can vary widely in their focus and

TABLE 5.1
Examples of Regulatory and Status and Trend Environmental Monitoring Programs and Tools

Program	Monitoring Tools
Regulatory Monitoring	
Clean Water Act permits (US), Fisheries Act regulations (Canada), Water Framework Directive (EU), German Drinking Water Ordinance	Effluent and water chemistry; acute and sublethal toxicity testing; fish, shellfish, and benthic invertebrate surveys
Marine and Inland Water Sediment Disposal programs (US), CEPA Ocean Disposal Regulations (Canada), London Convention-Dredged Material Assessment Framework (International)	Sediment chemistry; acute and sublethal toxicity testing
Superfund (CERCLA) and Resource Conservation and Recovery Act (US); German and Dutch Soil Protection Acts (Germany; the Netherlands); Thematic Strategy for Soil Protection (EU)	Soil and groundwater chemistry; aquatic and terrestrial toxicity tests (EU only)
OPP Pesticide Monitoring program (US); Pest Control Products Act regulation (Canada); Water Framework Directive and Emission Limit Values for Plant Protection Products (EU)	Tiered testing scheme: chemistry, appropriate nontarget testing, field trials
Status and Trend Monitoring	
NOAA National Status and Trends program (US); Environmental Monitoring and Assessment Program (US); National Shellfish Monitoring program (Canada); Water Framework Directive (EU)	Mussel watch monitoring; water, sediment, and tissue chemistry; sediment toxicity tests; benthic surveys
USGS Biomonitoring of Environmental Status and Trends (BEST) Program's nationwide large river monitoring network (US)	Fish tissue chemistry and health indicators
Arctic Monitoring and Assessment Program, Northern Contaminants Program (Canada)	Water, sediment, and tissue chemistry
Soil Indicators Consortium (UK)	Soil function and process measurements
ICES monitoring program for estuarine and near-shore environments and areas with offshore oil and gas operations (UK-CMAs, Nordic countries, Germany, the Netherlands)	Biota tissue analyses; waters and sediment chemistry; fish pathology and disease assessment; aquatic and sediment toxicity tests

monitoring activities (e.g., Herring gull egg shell analysis, forest pest collections after pesticide spray events, online monitoring stations in large rivers, freshwater and marine sediments, and fish tumor surveys). These programs are driven by the need to preserve health of specific resources such as shellfish or large regional ecosystems such as coastal areas, major river systems, and the Arctic (Long et al. 2003; Braune et al. 2005; Schmitt et al. 2005).

5.2.1 Current Approaches to Monitoring Exposure and Effects

Whether an assessment is driven by a regulatory or a status and trends monitoring program, evaluation of actual and potential contaminant exposure is performed after existing information about the site is used to identify the scope of the problem and assessment goals. Initial monitoring of exposure and effects falls into the second and third tiers in a general tiered assessment (Figure 5.1). The goal and purpose of the Tier II level is to provide a rapid screening level assessment of potential exposure to target stressors and potential adverse biological effects caused by suspect stressors.

This level of examination is generally low resolution in that the desire is to assess the potential for exposure and effects qualitatively. Current approaches at this level focus on determining levels of contaminant residues in environmental media or tissues and determining biomarkers of exposure in tissues. This can be done by laboratory or in situ toxicity tests to screen for toxicity (Figure 5.1). Contaminant residues in field-collected soil, sediment, water, or tissues are measured using standard analytical chemistry to determine contaminant bioavailability and associated risk posed to plants and animals. These techniques have played an important role in documenting the distribution and accumulation of persistent pollutants in fragile ecosystems such as the Canadian Arctic and specific populations such as the North American Great Lakes Herring gull (Hebert and Weseloh 2003; Braune et al. 2005).

5.2.2 Measurements of Biological Exposure and/or Effects

For many contaminants, mixtures of contaminants, or contaminants in combination with additional stressors, simple instrumental analysis of chemical concentration can sometimes fail to determine biological availability, accumulation, and/or the potential for adverse biological impacts and toxicity. Biological measurements currently used in monitoring include investigations at the molecular level, single cell, individual tissue, and the whole animal (for review, see Handy et al. 2003; van der Oost et al. 2003; Venturino et al. 2003; Galloway et al. 2004; Broeg et al. 2005; Tom and Auslander 2005). These measures, however, generally focus on a few endpoints at a time.

Ultimately, ecological risk assessments are concerned with monitoring effects that permit estimating potential impacts on population viability or success over several generations. Reproductive biomarkers for endocrine disruption currently in use include gene expression or immunoassays for vitellogenin levels in male fish (Lattier et al. 2001, 2002; Eidem et al. 2006), steroid hormones (estradiol, testosterone, and 11-ketotestosterone), histology of the gonads, annetocin—a reproductive biomarker in the earthworm, *Eisenia fetida* (Ricketts et al. 2004), and imposex effects in gastropods (Matthiessen and Gibbs 1998). These biomarkers are indicative of reproductive capacity and hence of impacts on population sustainability, but they have yet to gain wide acceptance in regulatory situations.

Exposure of animals to xenobiotics can alter their gene expression patterns, resulting in changes not only at the mRNA level, but at the protein and activity levels as well. While the changes at the mRNA level may not result in proportionate changes in protein concentration or activity, the changes themselves at any level of organization can provide some measure of exposure and effects. Examples of widely accepted protein biomarkers include metallothioneins 1 and 2 (Roesijadi 1994), amine oxidase, and the lysosomal associated glycoprotein, all of which have been proposed as early warning indicators of damage to *Lumbricus rubellus* when they are exposed to cadmium or copper (Burgos et al. 2005).

Another group of useful proteins includes enzymes involved in Phase 1 xenobiotic metabolism, which can provide chemical class specific biomarkers of exposure. For example, cytochrome P4501A1 (CYP1A1) (Stegeman and Lech 1991) is useful to determine exposure to polycyclic aromatic hydrocarbons (PAHs), polychlorinated biphenyls, polychlorinated dibenzo-p-dioxins, and furans. The activity itself can be measured by ethoxyresorufin-O-deethylase (EROD) activity (Levine and Oris 1997; Whyte et al. 2000; McClain et al. 2003; Roberts et al. 2005; Wilson et al. 2005). Since expression of CYP1A1 is controlled through the aryl hydrocarbon receptor, assays directly monitoring aryl hydrocarbon receptor and associated increased CYP1A1 protein levels are also being used (Carney et al. 2004; Aluru et al. 2005; Meyer et al. 2003; Gallagher 2006).

Metabolic markers are biochemical molecules created during normal or contaminant metabolism. This class of markers is useful to measure both direct and indirect effects of exposure. For example, one can measure changes in normal metabolite concentrations, such as sugars or creatinine, or in contaminant metabolites as a direct measure of impact. Alternatively, it is also possible to measure effects indirectly by examining adduction of bioactive metabolites to macromolecules including DNA and proteins. When combined with other measures such as enzymatic levels and pathology, metabolic markers can permit development of a relatively complete picture of chemical impact. For example, Weeks et al. (2004) found impacts of metals on soil invertebrates using a combination of metabolic, enzymatic activity, protein abundance, histopathological markers, and feeding behavior biomarkers.

Fish in Prince William Sound sites and the eastern Gulf of Alaska impacted by the 1989 *Exxon Valdez* oil spill were examined by monitoring bile fluorescent aromatic contaminants (FAC) and EROD activity (Huggett et al. 2003). The bile FAC results suggested that fish were exposed to low levels of PAH at all sites, even those remote to the spill, suggesting that the fish had been exposed to nonanthropogenic sources of PAHs. Winter et al. (2005) investigated effects of contamination in feral and caged chub (*Leuciscus cephalus*) from rivers in the West Midlands, United Kingdom, by examining bile metabolites of pollutants, hepatic biomarkers (EROD activity, levels of reduced glutathione, and serum aspartate aminotransferase).

5.2.3 Examples of Current Approaches to Exposure and Effects Monitoring

Status and trends monitoring programs such as those of the International Commission on Exploration of the Seas (ICES) and the United States Geological Survey (USGS) currently utilize an integrated combination of biomarkers; bioassays; tissue pathology; whole-animal health; and chemical analyses of biota, waters, and sediments. In an effort to monitor and prevent chemical impacts on coastal resources, ICES (2006) requires biological effects measurements to be made in estuarine and near-shore waters, shipping lanes, and areas with offshore oil and gas operations in the United Kingdom and other North Sea countries. Specifically, the currently accepted tests used for ICES assessments comprise acute toxicity tests of water and sediments (i.e., oyster embryo, the gammaridean amphipod *Corophium* and/or copepod *Tisbe battaglia* toxicity tests), markers of heavy metal and organic pesticide effects in mussels, imposex in gastropods, and biomarkers of exposure to polyaromatic hydrocarbons, metals, and endocrine disruptors.

The USGS program, Biomonitoring of Environmental Status and Trends (BEST), has established a Large River Monitoring Network (LRMN) to monitor the distribution of selected contaminants in fish tissue along with associated measures of health and biomarkers of exposure in resident fish species to elucidate linkages between exposure and biological impacts (Schmitt et al. 2005). Species selected for the LRMN have focused on a top predator species (typically largemouth bass) and a bottom-dwelling species (typically common carp). Biomarkers used in BEST are centered on functional systems, including reproductive systems, immune systems, and xenobiotic metabolism in the liver.

After gross field assessment of fish health is conducted for each fish, reproductive biomarkers are measured, including steroid hormones, vitellogenin, and histology of the gonads. Immune system measurements include quantification of macrophage aggregates in selected tissue and histology of major immune tissues. Liver function is evaluated through gross examination, histology, and EROD levels. Chemical measurements include screening for hydrophobic organic compounds (organochlorine pesticides and industrial chemicals) and a standard scan of metals, including mercury and selenium. Information for the LRMN is synthesized by river basin, with general inferences of contaminant impacts on fish populations within each collection area.

As we have noted, a number of tests and protocols have been developed and applied to assess exposure and effects in many different systems. However, current tests have several limitations—specifically, that only known endpoints are investigated and these may not be those that are impacted and/or most indicative of adverse effects. For example, several accepted current biomarkers are not dose responsive and may be induced at low-level concentrations that are tolerated by an organism, thus failing to predict impacts. Most currently used biomarkers are not specific for individual compounds but, rather, indicate exposure to broad classes of chemicals such as metals or estrogenic chemicals (Galloway et al. 2004). For many chemicals, there are no known biomarkers that can specifically indicate their presence.

Additionally it has been difficult to develop specific new assays for ecologically relevant species, especially at the molecular level since DNA sequence information is generally lacking for these species. Finally, existing assays do not deal well with complications that arise from interactive effects among stressors—for example, between a chemical exposure and seasonal variability. These limitations warrant exploration of alternate approaches to monitor exposure and effects.

5.2.4 Assessing Exposure and Effects with Mixtures of Contaminants

Current and historical activities by agriculture, industries, and governments have released a variety of chemicals into water bodies, sediments, and soils; the result has been plants and animals exposed to mixtures of chemicals. Whole effluent and whole sediment tests are routinely performed to address mixture effects in a regulatory context, but it is difficult to assess how individual components of the mixture are influencing the final outcome. Monosson (2005) identified several important shortcomings in mixture assessments: 1) few data points exist on many chemical mixtures; 2) of those that do exist, most evaluate only binary effects; and 3) few studies address chronic exposure to low environmentally relevant concentrations of chemicals. The difficulty in assessing real-world situations is illustrated by observations that sublethal concentrations of similar chemicals can act not only in a simple and predictable additive manner (Relyea 2004; Brian et al. 2005), but also in unpredictable ways such as by synergistic mechanisms that increase (Relyea 2003; Gust and Fleeger 2005; Relyea 2005) or antagonistic mechanisms that decrease apparent toxicity (Borgert et al. 2004; Gust and Fleeger 2006).

Understanding mechanisms of toxicity as well as having the ability to predict toxicological interactions among contaminants in chemical mixtures is fundamental to determining accurately the environmental impact of mixtures. The complex nature of mixtures can obscure the identity of the chemicals principally responsible for causing toxicity, even when the composition of the mixture is known. Toxicants in a mixture can, in principle, interact in 2 different ways: concentration addition or response addition (Altenburger et al. 2004).

The first method assumes that the components in a mixture act by the same mechanisms to cause similar effects and that addition of the respective concentrations of equivalent potency of each of the components will result in the observed response. The toxicity of the mixture therefore depends on the concentration of each component and their relative ratios, and assumes that the individual chemicals act jointly. For response addition, the components are believed to have dissimilar modes of action, but leading to the same overall effect. This second theory leads to calculation of a lower combined effect than one would get from concentration addition. However, both theories fail in very complex mixtures in which different groups of components exist, some of which act by similar modes of action and some of which do not (Altenburger et al. 2004).

The lack of understanding of how chemicals may interact with each other and biological systems greatly increases the uncertainty in estimating the risk posed by mixtures. Regulators often have used the default assumption that chemicals behave in an additive manner (Monosson 2005), and for the most part they do. An elegant example of additivity for compounds affecting the same biochemical pathway was shown for a mixture of estrogenic compounds including 2 strong estrogens (17-β-estradiol and ethinylestradiol) and 3 weak environmental estrogens (nonylphenol, octylphenol, and bisphenol), These were all shown to induce vitellogenin in male fathead minnow (Brian et al. 2005). Furthermore, when they were combined in equipotent concentrations, the effects of the mixture fit a mathematical model for additivity. For another mixture, however, synergy was used to explain the effects of mixing 2.2′,4,4′,5,5′-hexachlorobiphenyl and 2,3,7,8-tetrachlorodibenzo-p-dioxin on hepatic porphyrin levels in the rat (van Birgelen et al. 1996).

One approach to a better understanding of how components interact involves fractionation of complex contaminated media using different resins, chelators, solvents, or precipitations (Brack and Schirmer 2003; Heinin et al. 2004; Grote et al. 2005). The resulting extracts are individually tested for toxicity with a variety of bioassays assessing short-term survival, growth, or reproduction. This method of fractionation and bioassay monitoring has been termed toxicity identification evaluation (TIE) or, more generally, "bioassay-directed fractionation" and is frequently used to identify causative substances in environmental samples (Hankard et al. 2005).

In addition to toxicity tests, specific biomarkers of effects have also been used to assess exposure based upon loss of toxicity after fractionation, but this approach is hampered by directly linking a biomarker to toxicity. For example, in the case of Cyp1A1, the biomarker tracks exposure, rather than effect. The induction of this biomarker has not been linked robustly to toxicity. This approach is time consuming but works quite well and has been used in several studies to discover compounds that may lead to estrogenicity or androgenicity in complex effluents—for example, in paper mill effluents (Hewitt et al. 2003) or in sewage effluents in the United Kingdom (Sheahan et al. 2002). Ideally, assays of mixture toxicity would provide direct measurement of subtle effects that would enable identification of the contributions of different chemicals and permit determination of additive, antagonistic, or synergistic effects. Genomic approaches may contribute directly to TIE measurements and give clearer understanding of potential interactions among chemicals in mixtures (Chapter 6).

5.2.5 Current Challenges in Environmental Monitoring and Assessment

Several difficult issues present challenges for the practice of environmental monitoring. There are currently no wholly satisfying approaches to assess the effects of chronic exposure to low concentrations of pollutants, to identify multiple effects caused by single chemicals, to provide insight into the effects of chemical mixtures or multiple stressors, and, finally, to meet the challenge of translating monitoring data into population- and ecosystem-level impacts (Eggen et al. 2004). The most fundamental approach to meeting these challenges requires detailed knowledge of the mechanisms governing these phenomena that can be integrated with ecological principles (Escher and Hermans 2002; van Straalen 2003; Eggen et al. 2004; Snape et al. 2004).

Current environmental monitoring programs are employed for the express purpose of identifying stressors of concern and highlighting the need for process modification or remedial action. However, environmental monitoring data can rarely be used to determine direct cause and effect relationships between stressors and observed effects in the field. The level of resolution at which current methods of environmental monitoring operate makes them helpful for identifying environments that have been impacted, but they are not well suited for providing mechanistic information about how stressors cause deleterious effects; thus, they fall short of guaranteeing protection of natural environments. It is likely that the complexity will mask the presence of contaminants that lead to subtle effects in exposed wildlife. These difficulties contribute to uncertainties in environmental monitoring, making it hard to distinguish "normal" versus "abnormal" responses in wildlife populations from contaminated and uncontaminated sites. While there are no easy solutions to this central problem associated with any biomonitoring program, it is possible that the application of genomics methodologies may help when these are combined with the more traditional biological endpoints.

Other, more practical challenges arise in monitoring contaminated sites where the need to examine large areas or volumes of material and multiple contaminants must be balanced with expensive analytical costs and limited available capital. For example, harbors, ports, and commercial waterways must maintain navigability by dredging and disposing of sediment. The heterogeneous distribution of contaminants and large volumes of material involved in most dredging operations can result in overly conservative ecological risk assessments that may delay or prevent remedial activities. This is particularly true when insufficient data on bioavailability and toxicity result in the unnecessary treatment of sediments where contaminants actually pose no risk.

Some limitations of current environmental monitoring programs that result in poor resolution of contaminant effects include

- relatively low number of measurement endpoints (because of the high cost of field sampling and testing)
- difficulty in defining causation
- low sensitivity of many endpoints
- focus almost entirely on apical effects that could be translated into population responses (e.g., morbidity and reproductive failure resulting in reduced population size), which do not yield mechanistic data
- difficulty in extrapolating laboratory and in situ data to natural populations and communities
- inability to determine and distinguish adaptive responses from adverse effects

Even with these limitations and uncertainties, current practices have established guidelines for baseline ecological status, detection of changes or trends, provision of feedback on process control and operations, assessment of remedial actions, and enforcement of regulations. But, the guidelines are imperfect, at best, and could be improved by more knowledge. This knowledge can be greatly expanded by application of new approaches and tools developed in the area of genomics.

5.3 THE PROMISE OF GENOMICS IN MONITORING

Meeting current challenges in monitoring requires approaches able to provide information on a large number of biological conditions, sensitive enough to detect early indicators of change (good or bad), and specific enough to monitor the causal mechanisms of change. The science of genomics (genetics, transcriptomics, proteomics, and metabolomics) has rapidly developed to the point where it can be used to gain mechanistic insights in addition to providing many unique solutions and opportunities in risk assessment (Travis et al. 2003; Hood et al. 2004; Pennie et al. 2004; USEPA 2004; Chan and Theilade 2005; Corvi et al. 2006). Perhaps the greatest attraction of genomic technologies is the potential to provide an "open", unbiased assessment of the health status of an organism without the need for a priori knowledge of the conditions to which the organism may have been subjected. This contrasts with the more limited targeted approach currently used in monitoring. This open nature of genomics also helps to address the desire to detect subtle changes. This is particularly important because adverse health effects may not be immediately noticed or may take a generation or more to manifest at a phenotypic level.

Several scenarios by which genomics have potential for aiding in the monitoring of environmental impacts of toxicants are shown in Figure 5.1. The implementation of genomics in monitoring programs can provide rapid, sensitive screening tools to assess (but not confirm) the presence and impacts of stressors analogous to multibiomarker approaches (Handy et al. 2003; Galloway et al. 2004). Considering the current developing state of the science, realistic possibilities for initial use of genomics data in regulatory programs would be for complex contaminant mixture screening or for ranking to set priorities for follow-up confirmatory toxicity testing or impact assessment (Purohit et al. 2003). The focus should be on evaluating genomics data in the context of the integrated biological effects of the mixture and/or multiple stressors on impacted organisms, rather than developing specific profiles for chemicals in the mixture.

Genomics can also be integrated into a monitoring program to guide experimental design and hypothesis generation, and reduce the amount of animal use currently required for monitoring programs, since multiple biochemical pathways can be investigated in the same experiment. An area where genomics may also be useful is in the identification of toxicant exposure and potential adverse effects, providing information about the mechanisms and modes of action for classes of chemicals and providing a potential signature of toxicity (e.g., Hamadeh, Bushell, et al. 2002).

It may be possible to use genomic tools noninvasively by sampling biological fluids that contain products related to toxicity—for example, metabolites of the toxicants themselves and from specific biochemical pathways activated by these exposures (Griffin et al. 2002). The metabolites appear in saliva, urine, or scat (Li et al. 2004). These fluids, plasma, or core tissue samples may also contain protein biomarkers released from organs and cells that have been targeted by the exposures (Ellis et al. 2002; DePrimo et al. 2003). Skin cells and feather pulp may also contain signature expression profiles that depend on exposure to toxicants (Wong et al. 2004; Meucci and Arukwe 2005; Rees et al. 2005). Thus, the use of genomic tools also permits conducting assessments using small, noninvasive samples, which will allow repeat sampling of the same organism over time.

5.3.1 POTENTIAL CONTRIBUTIONS OF GENOMICS TO REGULATORY ASSESSMENT

Genomics can be useful in environmental monitoring for nearly all conventional monitoring programs. Parties or agencies monitoring environmental health status of air, soil, and water quality could use genomic approaches as part of a battery of monitoring techniques. These tools will be useful as supplementary information in a weight-of-evidence approach to monitoring. Genomics cannot replace currently accepted monitoring tools, but would provide all monitoring agencies with screening and stressor identification tools that can rapidly assess the potential impacts of environmental contaminants and help with prioritization. Thus, programs requiring monitoring whole effluent toxicity, air quality, contaminated sites, groundwater quality, and hazardous waste facilities, as well as status and trends monitoring, are potential beneficiaries of genomics (Table 5.1).

5.4 APPLICATION OF GENOMICS IN ENVIRONMENTAL MONITORING

Perhaps the most direct manner to introduce the many different contributions that genomics can make to environmental monitoring is to elaborate upon them as they would be used in a generic tiered assessment as diagrammed in Figure 5.1. Genomics can provide a valuable complement to existing monitoring methods used in tiered testing and weight-of-evidence approaches. Monitoring programs are generally organized around a multitiered framework focused on different levels of organization and detail.

Tiered approaches have long been promoted for standardizing environmental risk assessments and serve as a convenient context in which to describe how genomics can be integrated into monitoring and assessment practices. These approaches have the advantage that appropriate technologies are used depending on the question asked, which ultimately leads to more efficient decision making. Genomics can aid investigations at each of the different levels. The techniques can allow the identification of novel biomarkers and the development of assay batteries aimed at identifying adverse effects. Usage of genomics in this context is anticipated to be achieved in the short term. Indeed, examples of the identification of potential novel biomarkers for classes of toxicants are already emerging from genomic analyses.

5.4.1 USING GENOMICS TO AID IN PROBLEM FORMULATION AND HYPOTHESIS GENERATION: TIER I

The problem formulation stage sets up how the monitoring will progress, describes the problem that is being addressed, and develops the hypothesis to be tested. Problem formulation requires extensive review of existing data (historical, chemical, and environmental) to establish and identify potential exposure routes, relevant ecological species, and areas of concern. It is at this stage that one has to agree on the goals of the investigation and determine the appropriate thresholds of effects or measurements that will trigger successive tiers of evaluation. This is the stage where plans are set to monitor the effectiveness of the proposed actions and exit criteria are established. These formulations are essential for conducting effective ecological risk assessments. The most important step is to develop a good hypothesis, because this helps to focus the study and will help to formulate a framework for rigorous assessment and ultimately reduce unnecessary testing.

5.4.1.1 Genomics Can Assist in Problem Formulation and Hypothesis Generation in Real-World Situations: Examples from Several Current Studies

Genomics can provide substantial support for generating a testable hypothesis to understand and monitor the impact of chemicals on biological systems. Similar to tools in forensic sciences, genomics can screen for effects in organisms of concern. Effects can be observed in samples collected from animals in the field and tested for changes in pathways related to specific responses, such as hormonal pathways, damage repair, chemical specific exposure effects, disease pathology, reproductive effects, or general health status.

Transcriptomic, proteomic, and metabolomic methods for use in monitoring efforts are currently being explored as fundamental concepts and applications are being developed (Robertson 2005). These efforts have pointed out many potential uses for environmental monitoring. For example, nuclear magnetic resonance (NMR) analysis was used to measure fluoroanaline metabolites in earthworms exposed to 2 model xenobiotic compounds—4-fluoroaniline and 4-fluorobiphenyl—to determine the effects of Phase 1 and Phase 2 enzymes in gut microflora (Bundy et al. 2002). In another study, NMR was used to measure changes in metabolite profiles of earthworms exposed to metals (Bundy et al. 2004). Microarray analyses have been used in other studies to measure changes in gene expression profiles of fish exposed to endocrine disrupting compounds in laboratory experiments or in effluents (Larkin et al. 2003; Sheader et al. 2004; Wang et al. 2004). Genomics has also been used to establish the mechanism by which chromium inhibits transcription of genes that are sensitive to PAHs (Wei et al. 2004).

Examples of how genomics can assist in problem formulation and hypothesis generation in real-world mixtures are given by studies on effluent toxicity. Effluents from industrial processes contain complex mixtures of waste and byproducts that can potentially damage the environment. Chemical analysis of effluents does not always identify problematic releases. Results from exploratory genomic evaluation of effluents from pulp and paper mills identified potential areas of concern and helped in development of hypotheses to test during future monitoring or remediation efforts. In one study of an antiquated paper mill in Florida, differential display reverse transcriptase polymerase chain reaction analysis of largemouth bass liver gene expression showed that the effluents did not elicit estrogen-like activities. The effluents did, however, elicit alterations in gene expression patterns consistent with other modes of action, including the increased expression for Cyp1A and other as yet unidentified genes; this suggested exposure to aromatic toxicants (Denslow et al. 2004).

In a second study, effluents collected from 9 Canadian pulp and paper mills were monitored for sublethal effects using microarrays (van Aggelen 2005). Juvenile rainbow trout exposed to effluents in standard acute lethality 96-hour tests were analyzed with a 151-gene microarray. Due to the complexity of the effluent contaminant mixture, it was difficult to correlate chemical analysis and biological responses. However, observations that 6 effluents induced an estrogenic response and increased expression of detoxification genes coupled with down-regulation of stress genes suggested a significant chemical challenge to the exposed fish. The microarray data provided a more accurate qualitative indicator of responses to effluent exposure, demonstrating how genomic techniques can be used to enhance the information generated from routine compliance monitoring tools and focus assessments on specific stressors.

5.4.2 GENOMICS AND EXPOSURE AND EFFECTS SCREENING AND MONITORING: TIER II

As noted previously, once the assessment and monitoring problem has been defined, contaminated media are examined for potential to cause bioaccumulation of suspect chemicals in addition to adverse biological effects. Animals potentially exposed to contaminants on-site may also be examined. Genomics can provide many tools complementary to existing techniques to study contaminant impacts. These tools will be especially valuable if they are anchored to physiological endpoints or tissue architectural changes. For example, initial screening with microarrays could be used to develop more appropriate markers for specific chemicals or stressors.

5.4.2.1 Development of Screening Tools for Exposure and Effects Using Genomics

The most immediate application of genomics is in developing new screening assays of exposure and effects (see Chapter 2). The holistic nature of genomics makes it especially suited for biomarker discovery. The investigator is not limited to studying only anticipated genes or pathways that have been previously linked to a specific toxicant. Instead, all pathways that are affected by the toxicant will be unveiled and some of the genes in new pathways may prove to be ideal biomarkers of exposure because they provide specificity and/or predictive utility (in terms of apical responses). Since most contaminated sites include many chemical stressors, biomarkers that remain responsive and specific in the presence of other stressors will be necessary.

Genomic techniques will also allow for the discovery of biomarkers for species with limited sequence information. For example, the metallothionein genes in *Daphnia magna* were discovered using anonymous cDNA arrays (Poynton et al. 2006). Finally, it will be possible to utilize smaller and, more importantly, noninvasive sampling techniques. Samples such as saliva, blood, needle biopsies, surface mucus of fish, urine, skin cells collected with tape, nonlethal gill biopsies, and more have been used to perform genomic analysis (Ellis et al. 2002; Griffin et al. 2002; DePrimo et al. 2003; Li et al. 2004; Wong et al. 2004; Meucci and Arukwe 2005; Rees et al. 2005; Wang et al. 2005).

In developing new biomarkers of exposure and effect, dose- and time-dependent effects must be considered. Investigators can search for dose-responsive biomarkers affected at a threshold concentration known to cause an effect and eliminate biomarkers that are affected at lower concentrations or that may be influenced by the natural environment. For example, Perkins and Lotufo (2003) used differential display to identify several genes in the benthic amphipod *Leptocheirus plumulosus* whose expression correlated with sublethal tissue concentrations of 2,4,6-trinitrotoluene.

Screening level tools should also be tied to chemical bioaccumulation. Many pollutants have the potential to bioaccumulate and biomarkers that respond proportionately to chemical concentrations would be helpful to differentiate between long-term and transient exposure. As an example of such a biomarker, metallothionein expression has been successfully correlated with metal concentrations in invertebrate tissues (Mourgaud et al. 2002; Dallinger et al. 2004; Lecoeur et al. 2004).

Among the different genomic technologies, metabolomics may provide the most conserved measures of exposure and effect since metabolites are the common products of pathways in diverse species. Metabolite measurements are also the most difficult to obtain because of the quantities of product necessary for measurement. However, as new methods for microanalysis of metabolites are developed, this problem will be removed (Viant 2003; Collette et al. 2005; Robertson 2005).

Changes in protein levels may also be advantageous since many proteins share sequence homologies across species and antibodies developed against important proteins from a laboratory animal may cross-react with species from the field, as was illustrated by the universal antibody developed against rainbow trout (*Oncorhynchus mykiss*) vitellogenin (Heppell et al. 1995). Measurements made at the mRNA level will be the most difficult to extrapolate across species since there is considerable variation in nucleotide sequence (due to the degeneracy of the genetic code), even for genes that encode highly homologous proteins. Nevertheless, some studies show the utility of using microarrays among closely related species (Ji et al. 2004; Chalmers et al. 2005).

5.4.2.2 Biological Significance: Linking Changes at the Molecular Level with Adverse Effects

The ability to link genomic endpoints to an adverse biological effect is necessary for the widespread acceptance of these techniques. Genomic measurements may be more sensitive than traditional measures of toxicity, and many worry that they will be applied indiscriminately. A genomic alteration alone may simply reflect an adaptive or compensatory response and not actually indicate an adverse outcome. Molecular changes at the mRNA and protein levels need to be related to phenotypic changes in physiology and pathology that are known to predict adverse effects—called "phenotypic anchoring". The occurrence of ovo-testis in fish (Kirby et al. 2004) or mollusks (Horiguchi et al. 2000) are examples of tissue changes that may affect reproductive success and consequently population health.

A dimension that must be integrated into using genomics for environmental monitoring is the dependence of certain mechanisms on dose-dependent transitions, such that exposure to increasing concentrations of chemicals provokes increasingly severe consequences within an organism (Figure 5.2). It is relatively straightforward to connect molecular-level events to observable effects at the physiological level at high doses. However, in the real world, organisms are more likely to be exposed to a gradient of doses, with most exposures at low levels for long periods of time. In order to build a strong model relating molecular effects to more tangible higher order effects, we need a comprehensive dose–response and time-dependent knowledge base that includes molecular as well as more traditional endpoints. The importance of this approach has been shown in toxicogenomic studies that have successfully anchored gene expression profiles to pathological conditions in rats (Hamadeh, Knight, et al. 2002; Moggs et al. 2004). Translating their experimental design to ecotoxicology should allow investigators to meaningfully relate changes that occur at each level of biological organization over a range of environmentally relevant concentrations.

Perhaps the most important and most contentious issue in screening for effects is defining a "significant effect" in an often variable background. To define sig-

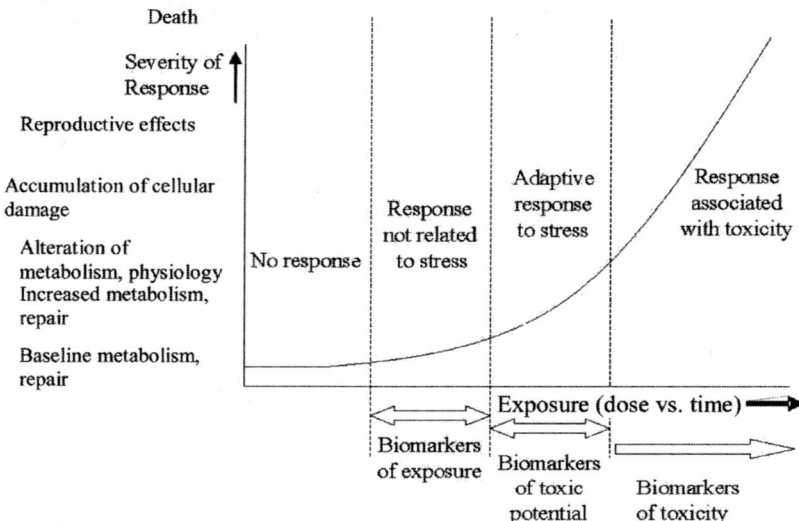

FIGURE 5.2 The linkage between genomic responses and adverse outcomes of concern to regulators. Note that this also is the area of greatest uncertainty—extrapolating results down to low doses. (Adapted from Klaper et al. 2003).

nificance, we must understand "natural" or "normal" background variability in gene, protein, or metabolomic expression profiles. Intrinsic sources of variability in expression profiles include different physiological states, age, sex, and genetic polymorphisms in natural populations. Connecting molecular changes in expression profiles to detrimental outcomes through phenotypic anchoring will allow investigators to ascertain significant changes.

Phenotypic anchoring will also aid in reducing uncertainty in safety factors. As the severity of the response being measured increases, the safety factor also increases in order to realize a "no-adverse-effect level". A relatively large degree of uncertainty is currently associated with large safety factors. However, if it is possible to link severe outcomes to certain molecular responses seen at lower concentrations of exposure, then the molecular responses could be monitored and would indicate the potential of the severe outcome. The importance of this approach was illustrated by Hamadeh, Knight, et al. (2002), whose study revealed that gene expression changes provided an early indication of heptatoxicity and were more sensitive than histopathologic endpoints. By improving detection of responses lower down on the dose–response curve (Figure 5.2), genomics screening tools will create more confidence in setting correct concentration safety levels and will provide a better understanding of the dose–response relationship.

5.4.3 Assessing Acute and Chronic Effects with Genomics: Tier III

Sites for which significant levels of exposure or effects have been established are currently examined in more detail in a third tier (Figure 5.1), with the objective of quantitatively defining the level of toxicity in acute and chronic toxicity tests.

Specifically, the potential for the contaminated soil, sediment, effluent, or water to cause toxicity is examined by various biological tests (EC 1999; USEPA 2000; USEPA 2002; Power and Boumphrey 2004; ASTM 2005). These toxicity tests permit screening for effects on survival, growth, and reproduction and relating effects to chemical bioaccumulation in standard organisms, preferably closely related to those inhabiting the monitored site (e.g., Markwiese et al. 2001; Achazi 2002; Hodkinson and Jackson 2005). Standardized acute, sublethal, and chronic toxicity tests (e.g., algal growth inhibition, invertebrate development, and fish lethality; fish embryo development and growth) are used in a regulatory context to monitor the cumulative toxicity of mixed contaminants in waste streams and aquatic environments (USEPA 1991; EC 1999; Johnson et al. 2004; Tinsley et al. 2004).

Genomics can complement standard approaches and provide far greater information to guide assessment of risk. More information can be captured from existing and developing protocols by monitoring impacts on multiple toxicological or stress response pathways, interpreting the significance of sublethal effects, and monitoring genomic indicators predictive of long-term effects. Genomic-based methods applied in weight-of-evidence approaches would improve the present monitoring programs.

Adaptive and compensatory changes represent different ways that an organism can react to acute or chronic exposures to toxicants. An adaptive change is more benign in that it allows an organism to tolerate a toxicant by detoxifying the toxicant or accelerating its elimination. A compensatory change may involve changes in gene expression or activity within the organism that help to counterbalance the effects of a toxicant—for example, the activation of a relatively unused biochemical pathway to take the place of one that has been damaged by the chemical action. Organisms can tolerate or adapt to chemical exposure by increasing production of protective mechanisms (e.g., CYP1A1, metallothioneins, heat shock proteins) or decreasing production of targets that confer sensitivity (e.g., mutation of genes for aryl hydrocarbon receptor; Okey et al. 2005).

Organisms may also compensate for the impairment of an essential function by changing a second function, an indirect effect of chemical exposure. Gene expression changes associated with defense mechanisms that are beneficial or adaptive may bear no causal relationship with the development of pathologies and must be separated from those that damage key cell functions (e.g., cell cycle control, structural integrity of proteins, control of free radicals, DNA repair mechanisms, and loss of homeostasis). It may be difficult to distinguish between a compensatory change in the expression profiles and adaptive changes that are actually associated with toxicity. Such concerns may require focusing on biological pathways with known adverse effects (e.g., endocrine disruption, DNA damage, apoptosis, or cell proliferation) rather than relying on single gene responses such as heat shock proteins.

A single chemical can affect multiple target sites, cause multiple effects, exhibit multiple modes of action, and have both time- and dose-dependent consequences that are particular to different species and developmental times. In order to understand how to apply genomics in mechanistic risk assessment, we need to understand which molecular events are most critical in the progression of biological effects, how each step relates to the next and feeds back on other steps, and their temporal relationships. Results must be interpreted within the context of multiple confounding factors such as time, seasonal variation, circadian cycles, reproductive status, chemical routes of exposure, and spatial heterogeneity in contaminant exposure. This is illustrated in observations that circadian rhythms have dramatic effects on gene expression observed in rat livers, thereby demonstrating the need to be aware of and to minimize confounding factors (Boorman et al. 2005). However, the same caveats apply for all biomarkers of physiological responses.

5.4.4 USING GENOMICS IN ASSESSMENTS ON COMPLEX, CASE-SPECIFIC CONDITIONS: TIER IV

Assessments often progress to the point where more information is needed than can be derived from simple exposure studies or acute and chronic effects tests. A fourth tier explores in detail the mechanisms behind the effects using nonstandard testing on a case- and problem-specific manner. The ability to identify disturbance of biological processes (pathways) through genomics takes the emphasis away from the need to design toxicity measurements that are specific for a particular chemical. In this section, mixture effects and interactions of multiple stressors, monitoring different species, and extrapolating genomic-level effects to population level impacts are examined.

5.4.4.1 Understanding Chemical Mixture Effects and Interactions of Multiple Stressors

Contaminants are seldom found in the environment alone and their toxicity in mixtures may be quite different from how they function individually. There is a lack of knowledge about how individual components of a mixture affect the uptake, biotransformation, toxicity, and excretion of other components in the mixture. Both additive and synergistic effects have been demonstrated, as discussed in Section 5.2.4.

Other stressors in the environment, such as increased UV radiation, hypoxia, nutrient loads, pathogens, temperature fluctuations, urban and agricultural runoffs, and other habitat quality issues, further complicate understanding toxicity. Even when the effluent being studied is predominantly from a single industry—for example, paper mill effluent—it is quickly apparent that different mills use a variety of different chemical processes to make paper and a variety of woods for furnish, such that each effluent is unique in time and place.

An example of the effects of mixed stressors is seen in a study by Relyea (2005), which showed that exposure of amphibian larvae to sublethal levels of roundup (glyphosate plus surfactant) in the presence of a stressor such as the chemical cues emitted by predatory newts caused increased tadpole mortality. In another study, Relyea (2004) showed that combinations of pesticides (diazinon, carbaryl, malathion, and glyphosate) caused additive effects on tadpole survival. The large number of possible combinations of toxicants plus stressors present in the environment precludes testing interactions under carefully controlled laboratory exposures. Examining effects on animals in the contaminated environments can help identify unusual or unexpected interactions and increase our understanding of adverse effects from mixture interactions.

Genomics using high-density microarrays lends itself to a broad assessment of chemical impacts of chemical mixtures and combinations of contaminants with other stressors. This approach has the potential to overcome the difficulties associated with reliance on biomarkers that may be elevated by one class of chemical and reduced by another. For example, induction of CYP1A1 has been used as a biomarker of exposure to polycyclic aromatic hydrocarbons, but the induction is inhibited (through metal response elements in the CYP1A1 promoter) by metals such as cadmium, which may be present in a complex environmental mixture (Lewis et al. 2004).

There may be other genes that are up-regulated by PAHs that do not have metal response elements and would continue to show response in the presence of metals. Only through a system-wide approach, such as is offered by genomics, will these be discovered. A complete understanding of the effects of mixtures in complex environmental settings will only arise out of understanding the mechanistic basis for the changes in growth, reproduction, and survival. This will require an integrated multi-tiered approach in which genomics technologies could play a major role.

5.4.4.2 Defining Modes and Mechanisms of Action Can Answer Many Questions

The regulatory arena makes a clear distinction between the terms "mode of action" and "mechanism of action". Mode of toxic action is a set of physiological, biochemical, and behavioral signs characterizing a specific biological effect (Borgert et al. 2004). Mechanism includes the mode of action but also requires understanding the exact way in which a toxicant affects a target to cause a response. The challenge for genomics will be to define the mode of action of a toxicant quickly and efficiently and provide an initial hypothesis of a mechanism of action that can be refined and defined by further research (USEPA 2003; Wei et al. 2004). The mechanism of action is built consecutively from a detailed knowledge of the mechanisms underlying the changes observed at the mRNA, protein, and activity levels (Escher and Hermens 2002; van Straalen 2003; Eggen et al. 2004; Snape et al. 2004).

The USEPA Office of Pesticides applies mechanistic data to determine whether the registration of pesticides should be based on risks posed by exposure to each individual pesticide or to an entire class of pesticides, such as organophosphates. Environmental monitoring and subsequent environmental risk assessments focus on mechanisms of action that begin with clearly defined toxicological interactions at the molecular level and cascade into predictable outcomes at the cellular, tissue, or organismal level. Recent research suggests that genomics can play a significant role in establishing mechanisms and modes of action. For example, the mechanism of hydrazine neurotoxicity in rats has been further elucidated using metabolomic analysis (Nicholls et al. 2001).

USEPA guidelines refer to the need for establishing the causal connections between mechanistic steps in order to characterize a chemical's mode of toxicity. It is clear that genome-wide alterations will be useful to identify signature patterns of gene expression changes that are involved as a consequence of exposure. But, to derive mechanistic information from these experiments will require additional methods of analysis, such as in situ hybridization, immunohistochemistry, and cell-specific quantitative-PCR. The most powerful application of genomics data may be when they are used in combination with current approaches in determining gene function, protein–protein interactions, knockdown mutants of specific genes, cellular and physiological targets of action, toxicokinetic profiles, parallel dose–response curves, and additive, synergistic, antagonistic interaction.

5.4.4.3 Determining Impacted Functions through Pathway Analysis

The functions of genes placed on microarrays can be identified by annotations using structured, controlled vocabularies and classifications covering several domains of molecular and cellular biology known as gene ontology (GO) terms (Harris et al. 2004). Changes in genes with related functions can be identified by testing for enrichment or overrepresentation of GO terms in the list of genes significantly affected by toxicant exposure (Khatri and Draghici 2005). Since GO terms can represent functions and associations with pathways, an enrichment of genes can help identify pathways affected by exposure. For example, gene ontology mapping was used as a method to reduce data from hundreds of individual genes to enable identification of molecular pathways and processes affected by the carcinogen diethylhexylphthalate and the various networks involved in responses to the anticarcinogen 3H-1, 2-dithiole-3-thione (Currie et al. 2005, Huang et al. 2006). If ontologies can be developed for protein and metabolite data sets, similar approaches may be useful in reducing those data in a meaningful way.

To understand the approach, it may be helpful to envisage a hypothetical example. Let us consider the possibility that genomic techniques and GO term interrogation were to identify an increase in expression of genes and proteins involved in DNA repair. This by itself is not an adverse effect and should not mandate immediate regulatory action. Instead, this type of result should be seen as an alert that would activate more detailed studies in Tier IV analysis (Figure 5.2). Such an approach would be a proactive strategy to avoid what could become a more severe outcome in exposed organisms if left to persist. Current strategies may not detect potential problems until too late.

Less simplistic might be the identification of disruption of multiple functional pathways that inform on the extent of tissue disturbance. As an example, induction of antioxidant systems and associated changes in the metabolome in isolation might signal potential exposure to reactive oxygen species in which the organism has adequately compensated to avoid toxicity. However, such changes in association with other, more severe pathway modulation such as apoptosis, coupled with targeted assessment of lipid peroxidation and pathological change, might point to the need for regulatory action. Such comprehensive findings, in which genomics contribute to identify the modes of action, will also assist greatly in the identification of the class of agent responsible. The preceding responses along with increased expression of biomarkers such as metallothionein and other metal-responsive genes might incriminate metal contaminants as likely causative agents.

5.4.4.4 Extrapolation of Chemical Effects across Many Species

The species that may be impacted by chemicals vary from site to site—from rainbow trout in streams and flounder in marine environments to deer mice in meadows and earthworms in soil. It is impossible to test every species with every stressor, due to the magnitude of the chemical stressor and species matrix. As a result, regulators are forced to extrapolate from effects on known species to those with unknown sensitivities. Genomics can help establish relationships between species through mechanisms of action, conservation of genetic pathways, and conservation of responses. Several techniques are available to build the resources needed to monitor different species, including: cross-species hybridizations, massive high-throughput sequencing, methods to create libraries of cDNAs representing different stressor effects or life stages, small molecule profiling with NMR, and increasingly more efficient methods to characterize proteins. Note the particular utility of metabolomics because of the uniformity of molecule identity between species.

A powerful strategy to compare species would be to establish networks representing interacting genes, proteins, and metabolites in model organisms (Luscombe et al. 2004; Dohr et al. 2005; Patil and Nielsen 2005; Villeneuve et al. 2006). Perturbation of these networks can associate clusters of genes, proteins, or metabolites with toxicological responses (Begley et al. 2002). Extrapolations between species may be achieved by mapping genes from the uncharacterized species to similar genes within the network topology of chemical responses in a known model species. This orthologous comparison should be useful in providing testable hypothesis to validate cross-species extrapolation.

5.4.4.5 Reducing Uncertainty in Population Impacts by Defining Distribution of Individual Sensitivities

A further important value of genomics is the ability, using high-throughput analysis of single nucleotide polymorphisms and transcriptional profiles, to determine potential genetic selection that may have occurred as a result of pollutant impact. For example, polymorphisms related to pollutant distribution have been seen with the *Fundulus heteroclitus* populations in New Bedford Harbor, Massachusetts (Greytak et al. 2005), in the Elizabeth River, Virginia (Meyer et al. 2005), and elsewhere (Wirgin and Waldman 2004). An understanding of the distribution of genetic polymorphisms related to key genes, such as the aryl hydrocarbon receptor, involved in the response to dioxin-like compounds will decrease uncertainty in the extrapolation of effects by permitting more accurate population modeling (Nacci et al. 2002; Hahn et al. 2004; Greytak 2005; Okey et al. 2005). The phenomena of differential responses between individuals within a population to a chemical or idiosyncratic toxicity has been recognized in the medical field and is the impetus for individualized medicine (Orphanides and Kimber 2003; Waring and Anderson 2005).

The variability in gene expression among individuals captured by genomics can be used to assess distribution of sensitivity or robustness. Oleksiak et al. (2005) examined the functional importance of individual variation by comparing cardiac metabolism in the teleost fish *F. heteroclitus* to patterns of mRNA expression. Both cardiac metabolism and metabolic gene expression were highly varied among individuals of natural outbred populations raised in a common environment. The variation in cardiac metabolism was found to be related to different patterns of gene expression in the different groups of individuals, suggesting that performance is related to variations in gene expression among individuals within a population. It seems reasonable that a similar approach using gene expression, protein or metabolite measures of variability anchored in phenotype or toxicological susceptibility could be used to estimate the distribution of responses within a population. The distribution of responses within a population could then be used in modeling population viability under different environmental conditions.

Genomic responses can be influenced by variability that may occur due to tissue specificity, circadian rhythms, age, and environmental conditions (Orphanides and Kimber 2003; Meyer et al. 2005; Waring and Anderson 2005). This problem is highlighted by a study of variability of background gene expression in mouse tissues, where genes associated with immune modulation, stress, and hormone responses varied up to 68 fold (Pritchard et al. 2001). The work of Pritchard and others highlights the need for more attention to be given to the extent of variability due to natural conditions in addition to modulation of biological pathways.

5.5 DEVELOPING GENOMICS FOR REGULATORY MONITORING

Organizations such as the EPA acknowledge that genomics technologies will eventually contribute to risk assessments and monitoring through a better understanding of mechanisms of toxicity, through identification of susceptible population, or for screening assessment (USEPA 2004). However, there is a need to develop acceptance criteria for genomic data. Indeed, a technical framework for analysis and acceptance criteria (e.g., data quality, experimental design) for genomics information for regulatory purposes needs to be addressed. A second challenge will be to reach common standards for methods, statistical analyses, and interpretations so that testing laboratories can perform genomic assays to meet regulatory requirements. While costs of equipment and material make application of current genomic approaches impractical for routine use, it is anticipated that these costs will decrease dramatically as technology improves and moves from a developing science to a more conventional approach (Ankley et al. 2006).

Validation of genomic tools will be a major task prior to their acceptance in routine environmental monitoring applications (Corvi et al. 2006). Genomic methods must be peer reviewed and validated in intra- and interlaboratory studies involving national or international standardization programs or organizations (e.g., USEPA, Environment Canada, American Society for Testing and Materials Standards, International Standards Organization, and Organization for Economic Co-operation and Development). This validation will define the degree of variability of the method and help establish the laboratory criteria (e.g., response in negative and positive controls, quality control) for validation of the method under controlled laboratory exposure conditions.

These validated genomic tools can then be applied in regulatory or status and trend field monitoring programs. Methods can be evaluated by using the tools in existing monitoring programs, as mentioned before, and in parallel to the standard suborganism or whole-organism parameters that are currently being measured. Eventually, validated genomic tools can be used to complement or even replace existing environmental monitoring tools. Part of the challenge will be in obtaining the participation of private sector consulting experts in the monitoring program, learning of genomic techniques by laboratories participating in monitoring, and, eventually, accreditation of laboratories who offer testing services (e.g., International Standards Organization Guide 17025 for accreditation of environmental testing laboratories). If the validation of the genomics tools is successful, laboratories should be accredited for use of new tools before any application in a regulatory monitoring program.

Education of regulatory and environmental staff who work for the regulated community will also be necessary for the use of genomics in regulatory programs. Stakeholders from industry, government, private consulting, and laboratory firms must be aware of why genomics tools are essential and useful. It should be noted that the use of sophisticated tools such as genomics will be incorporated into a regulatory monitoring program only if they are reliable (i.e., applied with sufficient quality and robustness), validated, and easily used and interpreted.

Opportunities exist within current programs to use genomics in the weight of evidence for impacts. Regulators in the United Kingdom have implemented a demonstration program examining the use of toxicity tests to manage and control effluents where mixed chemicals may occur (Tinsley et al. 2004; Wharfe et al. 2004). Similarly, strategies are currently being developed and implemented to incorporate genomics-based techniques into environmental monitoring programs in the United States (USEPA 2004). The USGS has collected, preserved, and archived liver, brain, and gonad samples from largemouth bass and common carp over the past 2 years of the BEST program for future genomic analysis of effects in large river systems.

The inclusion of more frequent sampling events within a long-term monitoring effort will greatly increase the likelihood of detecting subtle environmental trends (Hebert and Weseloh 2003). Given the expense and resource requirements associated with environmental sampling expeditions, the preservation and archiving of biological samples for future genomics analysis in addition to the standard sampling regime currently employed would provide a much greater "bang for the buck" in status and trends monitoring programs.

5.6 THE OUTLOOK FOR GENOMICS AND ENVIRONMENTAL MONITORING

5.6.1 Increasing Availability of Complete Genomes, Annotation, and Arrays

One of the necessary developments for increased use of genomic technologies in monitoring is the validation and commercialization of molecular tools for species normally evaluated in this context. Commercial assays will be indispensable for quality control and standardization of the methodologies among laboratories and, consequently, the validation of the method for monitoring of ecological areas. Currently, the limited number of commercially available arrays for the most important ecological species reduces the application of genomics in monitoring. However, there are several ongoing efforts around the world to commercialize such arrays. Commercial proteomic and metabolomic services for environmental models are also lacking, with the exception of a small set of commercially available antibodies for certain biomarkers, such as Cyp1A1, vitellogenin, metallothionein, heat shock proteins, and zona radiata proteins, among a few others.

There are many ongoing initiatives to sequence expressed genes for nonmodel species, but arrays for these models are mostly laboratory specific or include small numbers of genes, making the arrays useful for particular research efforts but not for whole-ecosystem monitoring. For example, the Daphnia Genome Consortium (http://daphnia.cgb.indiana.edu/) is a model collaborative effort to sequence the genome of an ecotoxicological organism and develop molecular tools, including cDNA and oligonucleotide microarrays, methods for cell line development and transfection, and transgenic organisms. As these initiatives continue to develop, the opportunity for using microarrays in ecosystem monitoring will increase.

The growing body of genomic information is making it possible to make inferences across multiple species and taxa, which will only enhance the resources available for ecosystem monitoring. There are a number of published studies suggesting that "homemade" arrays can work exceptionally well to help understand the effects of pollutants on focused pathways (Larkin et al. 2003; Edge et al. 2005; Sheader et al. 2006; Soetaert et al. 2006). For ecosystem health monitoring, it is vital to understand how toxicants and natural stressors affect survival, growth, and reproductive success of target environmental species.

Genomic applications are currently limited by sparse genetic, proteomic, and metabolomic data as well as limited bioinformatics available for many species of interest. Genome annotation is still an emergent need. Even commercial arrays for the best developed research models, such as human, mouse, and zebrafish, suffer from poor annotation, where many genes are unidentified. It is expected that ecological models will benefit from efforts expended in better annotation of the human and mouse genomes, since many unidentified genes are similar to mammalian counterparts. As major strides are made in this arena in the near future, it will be possible to better understand how specific contaminant classes affect biochemical pathways. Eventually, using this approach, it will be possible to identify critical responses to acute exposure and to extrapolate to effects from chronic exposure.

Genomics will become amenable for a more complete characterization of toxicant effects as analyses achieve higher throughput and become more economical. The advantages of this approach when compared to conventional testing of ecotoxicity based on gross pathology is that more subtle, early warnings of potential harm may be identified. Such potential may transform the way in which regulatory decisions are made since the techniques are likely to detect stress responses in the absence of actual whole-organism manifestations of toxicity. Multiple sublethal toxicological or stress responses can be assessed in acute and chronic exposures by using focused arrays, reverse transcriptase PCR assays, or other methods to monitor changes in important pathways such as cellular or DNA damage repair, apoptosis, mitochondrial respiration, oxidative stress response, or neurotoxicity, to suggest a few.

5.6.2 Development and Maintenance of Databases Support Application of Genomics to Monitoring

The enormity of the data generated from genomics technologies requires new methods of analyses and statistical algorithms. This challenge is being met head on by many researchers involved in bioinformatics and has, for example, led to establishing the minimum information about microarray experiment (MIAME) standards (Brazma et al. 2001). Complex data sets such as those derived from genomics need to be simplified in order to determine statistically significant differences between clean and impacted environments. This can be achieved, for example, by principal components analyses that can define the "normal" range of parameters and identify conditions that deviate. Even more important will be the construction of toxicogenomic databases, especially as a resource for cross-species comparisons. Included in these databases will be primary data for response to chemical exposures at the tissue and organ level. One such database being constructed at the National Toxicology Program, National Institute of Health, is the Chemical Effects in Biological Systems. While this database will concentrate on mammalian models, there will also be information for lower vertebrates.

For monitoring paradigms it will be important to include information for species in the environment. Other groups have also come together to pool resources. For example, the Mount Desert Island Biological Laboratory has started a Comparative Toxicogenomics Database that should enable researchers to have access to work done on marine species and to compare their results easily with results obtained with mammals and humans. For fish that are not model species and therefore have limited genomic information available, it will be important to generate consortiums of interested researchers to obtain enough information to utilize these new technologies. However, annotation of genes found in nonmammalian species will continue to be a challenge for years to come.

5.6.3 THE FUTURE OF GENOMICS IN UNDERSTANDING EFFECTS: SYSTEMS TOXICOLOGY

Computational biology approaches will be increasingly important in understanding the responses at several different levels of biological organization. At the present time algorithms are focusing on changes in gene expression at the cellular or organ level. But to make this useful, we will have to develop algorithms that can extend the information to whole organisms, to a level of complex biological organization. While we think in linear sequences of events for cellular and tissue responses, it is possible that whole organisms will show nonlinear responses to environmental stimuli (Hughes et al. 2000; Begley et al. 2002). We will need to include information in our models for complex adaptive, pharmacological, and toxicological responses. The models will have to reflect complex kinetic and dynamic changes in specific tissues. Most probably we will have to develop new algorithms that use complex network behavior and nonlinear mathematical models. This new approach is termed systems toxicology or computational toxicology (Figure 5.3) and refers to understanding global changes in organisms in response to toxicants and other stressors (Hood et al. 2004; Waters and Fostel 2004; Villenneuve et al. 2006).

Arguably the biggest challenge in genomics will be in extrapolating effects from the molecular scale to the cellular level to individuals to populations and, finally, to various species. This is a problem shared with existing ecological risk assessments where effects on cells and individuals must be extrapolated to populations. It is likely that as systems toxicology is fully developed, genomic technologies and associated

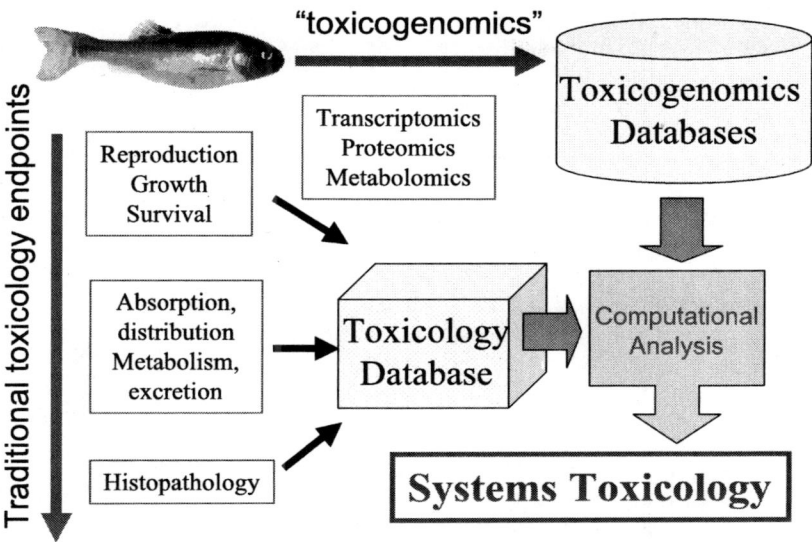

FIGURE 5.3 Schematic of systems toxicology. (Adapted from Waters and Fostel, 2004.)

computational analyses will become effective in identifying the health status of individuals and then whole populations. Systems toxicology will be able to integrate detailed mechanistic information from the molecular, cellular, and organism level to population and species responses within an ecosystem (Said 2004; Waters and Fostel 2004).

It is expected that systems toxicology will contribute to a more holistic approach for assessing the effects of mixtures by allowing scientists to begin to unravel specific effects of each individual chemical. It is expected that each contaminant in a mixture will interact with the biological system as a whole and also interact with the other contaminants to enhance or temper the overall effect as agonists or antagonists of specific pathways. A whole-systems approach to the problem, with sophisticated bioinformatics support, may allow scientists to tease out the specific effects of each contaminant to determine how each perturbs the system and how they interact when in combination.

REFERENCES

Achazi RK. 2002. Invertebrates in risk assessment: development of a test battery and of short term biotests for ecological risk assessment of soil. J Soils Sed 2:174–178.

Altenburger R, Walter H, Grote M. 2004. What contributes to the combined effect of a complex mixture? Environ Sci Technol 38:6353–6362.

Aluru N, Vuori K, Vijayan MM. 2005. Modulation of Ah receptor and CYP1A1 expression by alpha-naphthoflavone in rainbow trout hepatocytes. Compar Biochem Physiol Toxicol Pharmacol 141:40–49.

[ASTM] American Society for Testing and Materials. 2005. Biological effects and environmental fate; biotechnology annual book of ASTM standards, vol. 11.06. Philadelphia (PA): ASTM.

Ankley GT, Daston GP, Degitz SJ, Denslow ND, Hoke RA, Kennedy SW, Miracle AL, Perkins EJ, Snape J, Tillitt DE, et al. 2006. Toxicogenomics in regulatory ecotoxicology. Environ Sci Technol 40:4055–4065.

Begley TJ, Rosenbach AS, Ideker T, Samson LD. 2002. Damage recovery pathways in *Saccharomyces cerevisiae* revealed by genomic phenotyping and interactome mapping. Mol Cancer Res 1:103–112.

Boorman GA, Blackshear PE, Parker JS, Lobenhofer EK, Malarkey DE, Vallant MK, Gerken DK, Irwin RD. 2005. Hepatic gene expression changes throughout the day in the Fischer rat: implications for toxicogenomic experiments. Toxicol Sci 86:185–193.

Borgert CJ, Quill TF, McCarty LS, Mason AM. 2004. Can mode of action predict mixture toxicity for risk assessment? Toxicol Appl Pharmacol 201:85–96.

Brack W, Schirmer K. 2003. Effect-directed identification of oxygen and sulfur heterocycles as major polycyclic aromatic cytochrome P4501A-inducers in a contaminated sediment. Environ Sci Technol 37:3062–3070.

Braune BM, Outridge PM, Fisk AT, Muir DC, Helm PA, Hobbs K, Hoekstra PF, Kuzyk ZA, Kwan M, Letcher RJ, et al. 2005. Persistent organic pollutants and mercury in marine biota of the Canadian Arctic: an overview of spatial and temporal trends. Sci Total Environ 50:106–118.

Brazma A, Hingamp P, Quackenbush J, Sherlock G, Spellman P, Stoeckert C, Aach J, Ansorge W, Ball CA, Causton HC, et al. 2001. Minimum information about a microarray experiment (MIAME)—toward standards for microarray data. Nat Genet 29:365–371.

Brian J, Harris CA, Scholze M, Backhaus T, Booy P, Lamoree M, Pojana G, Jonkers N, Runnalls T, Bonfà A, et al. 2005. Accurate prediction of the response of freshwater fish to a mixture of estrogenic chemicals. Environ Health Perspect 113:723–728.

Broeg K, Westernhagen HV, Zander S, Korting W, Koehler A. 2005. The "bioeffect assessment index" (BAI). A concept for the quantification of effects of marine pollution by an integrated biomarker approach. Mar Pollut Bull 50:495–503.

Bundy JG, Lenz EM, Bailey NJ, Gavaghan CL, Svendsen C, Spurgeon D, Hankard PK, Osborn D, Weeks JM, Trauger SA, et al. 2002. Metabonomic assessment of toxicity of 4-fluoroaniline, 3,5-difluoroaniline and 2-fluoro-4-methylaniline to the earthworm *Eisenia veneta* (Rosa): identification of new endogenous biomarkers. Environ Toxicol Chem 21:1966–1972.

Bundy JG, Spurgeon DJ, Svendsen C, Hankard PK, Weeks JM, Osborn D, Lindon JC, Nicholson JK. 2004. Environmental metabonomics: applying combination biomarker analysis in earthworms at a metal contaminated site. Ecotoxicology 13:797–806.

Burgos MG, Winters C, Sturzenbaum SR, Randerson PF, Kille P, Morgan AJ. 2005. Cu and Cd effects on the earthworm *Lumbricus rubellus* in the laboratory: multivariate statistical analysis of relationships between exposure, biomarkers, and ecologically relevant parameters. Environ Sci Technol 39:1757–1763.

Carney SA, Peterson RE, Heideman W. 2004. 2,3,7,8-Tetrachlorodibenzo-p-dioxin activation of the aryl hydrocarbon receptor/aryl hydrocarbon receptor nuclear translocator pathway causes developmental toxicity through a CYP1A-independent mechanism in zebrafish. Mol Pharmacol 66:512–521.

Chalmers AD, Goldstone K, Smith JC, Gilchrist M, Amaya E, Papalopulu N. 2005. A *Xenopus tropicalis* oligonucleotide microarray works across species using RNA from *Xenopus laevis*. Mech Dev 122:355–363.

Chan VS, Theilade MD. 2005. The use of toxicogenomic data in risk assessment: a regulatory perspective. Clin Toxicol 43:121–126.

Collette T, Ekman D, Kenneke J, Whitehead T, Villeneuve D, Kahl M, Jensen K, Ankley G. Metabolomics as a diagnostic tool for small fish toxicology research. SETAC North America 26th Annual Meeting, Baltimore, MD, 13–17 Nov., 2005.

Corvi R, Ahr HJ, Albertini S, Blakey DH, Clerici L, Coecke S, Douglas GR, Gribaldo L, Groten JP, Haase B, et al. 2006. Meeting report: validation of toxicogenomics-based test systems: ECVAM-ICCVAM/NICEATM considerations for regulatory use. Environ Health Perspect 114:420–429.

Currie RA, Bombail V, Oliver JD, Moore DJ, Lim FL, Gwilliam V, Kimber I, Chipman K, Moggs JG, Orphanides G. 2005. Gene ontology mapping as an unbiased method for identifying molecular pathways and processes affected by toxicant exposure: application to acute effects caused by the rodent nongenotoxic carcinogen diethylhexylphthalate. Toxicol Sci 86:453–469.

Dallinger R, Lagg B, Egg M, Schipflinger R, Chabicovsky M. 2004. Cd accumulation and Cd–metallothionein as a biomarker in *Cepaea hortensis* (Helicidae, Pulmonata) from laboratory exposure and metal-polluted habitats. Ecotoxicology 13:757–772.

Denslow ND, Kocerha J, Sepulveda MS, Gross T, Holm SE. 2004. Gene expression fingerprints of largemouth bass (*Micropterus salmoides*) exposed to pulp and paper mill effluents. Mutat Res 552:19–34.

DePrimo SE, Wong LM, Khatry DB, Nicholas SL, Manning WC, Smolich BD, O'Farrell AM, Cherrington JM. 2003. Expression profiling of blood samples from an SU5416 Phase III metastatic colorectal cancer clinical trial: a novel strategy for biomarker identification. BMC Cancer 3:3–15.

Dohr S, Klingenhoff A, Maier H, Hrabe de Angelis M, Werner T, Schneider R. 2005. Linking disease-associated genes to regulatory networks via promoter organization. Nucleic Acids Res 33:864–872.

Edge SE, Morgan MB, Gleason DF, Snell TW. 2005. Development of a coral cDNA array to examine gene expression profiles in *Montastraea faveolata* exposed to environmental stress. Mar Pollut Bull 51:507–523.

Eggen RIL, Behra R, Burkhardt-Holm P, Escher BI, Schweigert N. 2004. Challenges in ecotoxicology. Environ Sci Technol 38:58A–64A.

Eidem JK, Kleivdal H, Kroll K, Denslow N, van Aerle R, Tyler C, Panter G, Hutchinson T, Goksoyr A. 2006. Development and validation of a direct homologous quantitative sandwich ELISA for fathead minnow (*Pimephales promelas*) vitellogenin. Aquat Toxicol 78:202–206.

Ellis M, Davis N, Coop A, Liu M, Schumaker L, Lee RY, Srikanchana R, Russell CG, Singh B, Miller WR, et al. 2002. Development and validation of a method for using breast core needle biopsies for gene expression microarray analyses. Clin Cancer Res 8:1155–1166.

[EC] Environment Canada. 1999. Guidance document on application and interpretation of single-species tests in environmental toxicology. Environmental Protection Series, EPS 1/RM/34, December 1999.

Escher BI, Hermens JL. 2002. Modes of action in ecotoxicology: Their role in body burdens, species sensitivity, QSARs, and mixture effects. Environ Sci Technol 36:4201–4217.

Gallagher EP. 2006. Using salmonid microarrays to understand the dietary modulation of carcinogenesis in rainbow trout. Toxicol Sci 90:1–4.

Galloway T, Brown R, Browne MA, Dissanayake A, Lowe D, Jones MB, DePledge MH. 2004. A multibiomarker approach to environmental assessment. Environ Sci Technol 38:1723–1731.

Greytak SR, Champlin D, Callard GV. 2005. Isolation and characterization of two cytochrome P450 aromatase forms in killifish (*Fundulus heteroclitus*): differential expression in fish from polluted and unpolluted environments. Aquat Toxicol 71:371–389.

Griffin JL, Nicholls AW, Keun HC, Mortishire-Smith RJ, Nicholson JK, Kuehn T. 2002. Metabolic profiling of rodent biological fluids via 1H NMR spectroscopy using a 1-mm microliter probe. Analyst 127:582–584.

Grote M, Brack W, Altenburger R. 2005. Identification of toxicants from marine sediment using effect-directed analysis. Environ Toxicol 20:475–486.

Gust KA, Fleeger JW. 2005. Exposure-related effects on Cd bioaccumulation explain toxicity of Cd-phenanthrene mixtures in *Hyalella azteca*. Environ Toxicol Chem 24:2918–2926.

Gust KA, Fleeger JW. 2006. Exposure to cadmium-phenanthrene mixtures elicits complex toxic responses in the freshwater tubificid oligochaete, *Ilyodrilus templetoni*. Arch Environ Contam Toxicol 51:54–60.

Hahn ME, Karchner SI, Franks DG, Merson RR. 2004. Aryl hydrocarbon receptor polymorphisms and dioxin resistance in Atlantic killifish (*Fundulus heteroclitus*). Pharmacogenetics 14:131–143.

Hamadeh HK, Bushel PR, Jayadev S, Martin K, DiSorbo O, Sieber S, Bennett L, Tennant R, Stoll R, Barrett JC, et al. 2002. Gene expression analysis reveals chemical-specific profiles. Toxicol Sci 67:219–231.

Hamadeh HK, Knight BL, Haugen AC, Sieber S, Amin RP, Bushel PR, Stoll R, Blanchard K, Jayadev S, Tennant RW, et al. 2002. Methapyrilene toxicity: anchorage of pathologic observations to gene expression alterations. Toxicol Pathol 30:470–482.

Handy RD, Galloway TS, DePledge MH. 2003. A proposal for the use of biomarkers for the assessment of chronic pollution and it regulatory toxicology. Ecotoxicology 12:332–343.

Hankard PK, Bundy JG, Spurgeon DJ, Weeks JM, Wright J, Weinberg C, Svendsen C. 2005. Establishing principal soil quality parameters influencing earthworms in urban soils using bioassays. Environ Pollut 133:199–211.

Harris MA, Clark J, Ireland A, Lomax J, Ashburner M, Foulger R, Eilbeck K, Lewis S, Marshall B, Mungall C, et al. 2004. The Gene Ontology (GO) Database and informatics resource. Nucleic Acids Res 32(database issue):D258–261.

Hebert CE, Weseloh DV. 2003. Assessing temporal trends in contaminants from long-term avian monitoring programs: the influence of sampling frequency. Ecotoxicology 12:141–151.

Heinin LJ, Highland TL, Mount DR. 2004. Method for testing the aquatic toxicity of sediment extracts for use in identifying organic toxicants in sediments. Environ Sci Technol 38:6256–6262.

Heppel SA, Denslow ND, Folmar LC, Sullivan CV. 1995. Universal assay of vitellogenin as a biomarker for environmental estrogens. Environ Health Perspect 103:9–15.

Hewitt LM, Pryce AC, Parrott JL, Marlatt V, Wood C, Oakes K, Van Der Kraak GJ. 2003. Accumulation of ligands for aryl hydrocarbon and sex steroid receptors in fish exposed to treated effluent from a bleached sulfite/groundwood pulp and paper mill. Environ Toxicol Chem 22:2890–2897.

Hodkinson ID, Jackson JK. 2005. Terrestrial and aquatic invertebrates as bioindicators for environmental monitoring, with particular reference to mountain ecosystems. Environ Manage 35:649–666.

Hood L, Heath JR, Phelps ME, Lin B. 2004. Systems biology and new technologies enable predictive and preventative medicine. Science 306:640–643.

Horiguchi T, Takiguchi N, Cho HS, Kojima M, Kaya M, Shiraishi H, Morita M, Hirose H, Shimizu M. 2000. Ovo-testis and disturbed reproductive cycle in the giant abalone, *Haliotis madaka*: possible linkage with organotin contamination in a site of population decline. Mar Environ Res 50:223–229.

Huang Y, Yan J, Lubet R, Kensler TW, Sutter TR. 2006. Identification of novel transcriptional networks in response to treatment with the anticarcinogen 3H-1, 2-dithiole-3-thione. Physiol Genomics 24:144–153.

Huggett RJ, Stegeman JJ, Page DS, Parker KR, Woodin B, Brown JS. 2003. Biomarkers in fish from Prince William Sound and the Gulf of Alaska: 1999–2000. Environ Sci Technol 37:4043–4051.

Hughes TR, Marton MJ, Jones AR, Roberts CJ, Stoughton R, Armour CD, Bennett HA, Coffey E, Dai H, He YD, et al. 2000. Functional discovery via a compendium of expression profiles. Cell 102:109–126.

[ICES] International Council for the Exploration of the Seas. 2006. Report of the second ICES/OSPAR workshop on integrated monitoring of contaminants and their effects in coastal and open-sea areas (WKIMON II). ICES Advisory Committee on the Marine Environment, ICES CM 2006/ACME: 02, 157 p.

Ji W, Zhou W, Gregg K, Yu N, Davis S, Davis S. 2004. A method for cross-species gene expression analysis with high-density oligonucleotide arrays. Nucleic Acids Res 32:e93.

Johnson I, Hutchings M, Benstead R, Thain J, Whitehouse P. 2004. Bioassay selection, experimental design and quality control/assurance for use in effluent assessment and control. Ecotoxicology 13:437–447.

Khatri P, Draghici S. 2005. Ontological analysis of gene expression data: current tools, limitations, and open problems. Bioinformatics 21:3587–3595.

Kirby MF, Allen YT, Dyer RA, Feist SW, Katsiadaki I, Matthiessen P, Scott AP, Smith A, Stentiford GD, Thain JE, et al. 2004. Surveys of plasma vitellogenin and intersex in male flounder (*Platichthys flesus*) as measures of endocrine disruption by estrogenic contamination in United Kingdom estuaries: temporal trends, 1996 to 2001. Environ Toxicol Chem 23:748–758.

Larkin P, Folmar LC, Hemmer MJ, Poston AJ, Denslow ND. 2003. Expression profiling of estrogenic compounds using a sheepshead minnow cDNA macroarray. Environ Health Perspec ToxicoGenomics 111:29–36.

Lattier DL, Gordon DA, Burks DJ, Toth GP. 2001. Vitellogenin gene transcription: a relative quantitative exposure indicator of environmental estrogens. Environ Toxicol Chem 20:1979–1985.

Lattier DL, Reddy TV, Gordon DA, Lazorchak JM, Smith ME, Williams DE, Wiechman B, Flick RW, Miracle AL, Toth GP. 2002. 17Alpha-ethynylestradiol-induced vitellogenin gene transcription quantified in livers of adult males, larvae, and gills of fathead minnows (*Pimephales promelas*). Environ Toxicol Chem 21:2385–2393.

Lecoeur S, Videmann B, Berny PH. 2004. Evaluation of metallothionein as a biomarker of single and combined Cd/Cu exposure in *Dreissena polymorpha*. Environ Res 94:184–191.

Levine SL, Oris JT. 1997. Induction of CYP1A mRNA and catalytic activity in gizzard shad (*Dorosoma cepedianum*) after waterborne exposure to benzo[*a*]pyrene. Comp Biochem Physiol C Pharmacol Toxicol Endocrinol 118:397–404.

Lewis N, Williams TD, Chipman K. 2004. Functional analysis of xenobiotic response elements (XREs) in CYP 1A of the European flounder (*Platichthys flesus*). Mar Environ Res 58:101–105.

Li Y, Zhou X, St John MA, Wong DT. 2004. RNA profiling of cell-free saliva using microarray technology. J Dent Res 83:199–203.

Long E, Dutch M, Aasen S, Welch K, Hameedi MJ. 2003. Chemical contamination, acute toxicity in laboratory tests, and benthic impacts in sediments of Puget Sound: A summary of results of the joint 1997–1999 Ecology/NOAA survey. Washington State Dept. of Ecology, Publication No. 03-03-049 and NOAA Technical Memorandum NOS NCCOS CCMA No. 163. Washington State Department of Ecology, Olympia, WA. http://www.ecy.wa.gov/biblio/0303049.html

Luscombe NM, Babu MM, Yu H, Snyder M, Teichmann SA, Gerstein M. 2004. Genomic analysis of regulatory network dynamics reveals large topological changes. Nature 431:308–312.

Markwiese JT, Ryti RT, Hooten MM, Michael DI, Hlohowskyj I. 2001. Toxicity bioassays for ecological risk assessment in arid and semiarid ecosystems. Rev Environ Contam Toxicol 168:43–98.

Matthiessen P, Gibbs PE. 1998. Critical appraisal of the evidence for tributyltin-mediated endocrine disruption in mollusks. Environ Toxicol Chem 17:37–43.

McClain JS, Oris JT, Burton GA Jr, Lattier D. 2003. Laboratory and field validation of multiple molecular biomarkers of contaminant exposure in rainbow trout (*Oncorhynchus mykiss*). Environ Toxicol Chem 22:361–370.

Meucci V, Arukwe A. 2005. Detection of vitellogenin and zona radiata protein expressions in surface mucus of immature juvenile Atlantic salmon (*Salmo salar*) exposed to waterborne nonylphenol. Aquatic Toxicol 73:1–10.

Meyer JN, Volz DC, Freedman JH, Di Giulio RT. 2005. Differential display of hepatic mRNA from killifish (*Fundulus heteroclitus*) inhabiting a Superfund estuary. Aquat Toxicol 73:327–341.

Meyer JN, Wassenberg DM, Karchner SI, Hahn ME, Di Giulio RT. 2003. Expression and inducibility of aryl hydrocarbon receptor pathway genes in wild-caught killifish (*Fundulus heteroclitus*) with different contaminant-exposure histories. Env Toxicol Chem 22: 2337–2343.

Moggs JG, Tinwell H, Spurway T, Chang HS, Pate I, Lim FL, Moore DJ, Soames A, Stuckey R, Currie R, et al. 2004. Phenotypic anchoring of gene expression changes during estrogen-induced uterine growth. Environ Health Perspect 112:1589–1606.

Monosson E. 2005. Chemical mixtures: considering the evolution of toxicology and chemical assessment. Environ Health Perspect 113:383–390.

Mourgaud Y, Martinez AR, Geffard A, Andral B, Stanisiere JY, Amiard JC. 2002. Metallothionein concentrations in the mussel *Mytilus galloprovincialis* as a biomarker of response to metal contamination: validation in the field. Biomarkers 7:476–490.

Nacci DE, Champlin D, Coiro L, McKinney R, Jayaraman S. 2002. Predicting the occurrence of genetic adaptation to dioxinlike compounds in populations of the estuarine fish *Fundulus heteroclitus*. Environ Toxicol Chem 21:1525–1532.

Nicholls AW, Holmes E, Lindon JC, Shockcor JP, Farrant RD, Haselden JN, Damment SJ, Waterfield CJ, Nicholson JK. 2001. Metabolomic investigations into hydrazine toxicity in the rat. Chem Res Toxicol 14:975–987.

Okey AB, Franc MA, Moffat ID, Tijet N, Boutros PC, Korkalainen M, Tuomisto J, Pohjanvirta R. 2005. Toxicological implications of polymorphisms in receptors for xenobiotic chemicals: The case of the aryl hydrocarbon receptor. Toxicol Appl Pharmacol 207:43–51.

Oleksiak MF, Roach JL, Crawford DL. 2005. Natural variation in cardiac metabolism and gene expression in *Fundulus heteroclitus*. Nat Genet 37:67–72.

Orphanides G, Kimber I. 2003. Toxicogenetics: applications and opportunities. Toxicol Sci 75:1–6.

Patil KR, Nielsen J. 2005. Uncovering transcriptional regulation of metabolism by using metabolic network topology. Proc Natl Acad Sci USA 102:2685–2689.

Pennie W, Pettit SD, Lord PG. 2004. Toxicogenomics in risk assessment: an overview of an HESI collaborative research program. Environ Health Perspect 112:417–419.

Perkins E, Lotufo G. 2003. Playing in the mud—using gene expression to assess contaminant effects on sediment dwelling invertebrates. Ecotoxicology 12:453–456.

Power E, Boumphrey R. 2004. International trends in bioassay use for effluent management. Ecotoxicology 13:377–398.

Poynton HC, Varshavsky JR, Chang B, Cavigiolio G, Chan S, Holman PS, Loguinov AV, Bauer DJ, Colbourne JK, Komachi K, et al. 2007. *Daphnia magna* ecotoxicogenomics provides mechanistic insights into metal toxicity. Environ Sci Technol 41:1044–1050.

Pritchard CC, Hsu L, Delrow J, Nelson PS. 2001. Project normal: defining normal variance in mouse gene expression. Proc Natl Acad Sci USA 98:13266–13271.

Purohit HJ, Raje DV, Kapley A, Padmanabhan P, Singh RN. 2003. Genomics tools in environmental impact assessment. Environ Sci Technol 37:356A–363A.

Rees CB, McCormick SD, Li W. 2005. A nonlethal method to estimate CYP1A expression in laboratory and wild Atlantic salmon (*Salmo salar*). Comp Biochem Physiol C Toxicol Pharmacol 141:217–224.

Relyea RA. 2003. Predator cues and pesticides: a double dose of danger for amphibians. Ecol Appl 13:1515–1521.

Relyea RA. 2004. Growth and survival of five amphibian species exposed to combinations of pesticides. Environ Toxicol Chem 23:1737–1742.

Relyea RA. 2005. The lethal impacts of roundup and predatory stress on six species of North American tadpoles. Arch Environ Contam Toxicol 48:351–357.

Ricketts HJ, Morgan AJ, Spurgeon DJ, Kille P. 2004. Measurement of annetocin gene expression: a new reproductive biomarker in earthworm ecotoxicology. Ecotox Environ Safety 57:4–10.

Roberts AP, Oris, JT, Burton GA Jr, Clements WH. 2005. Gene expression in caged fish as a first-tier method for exposure assessment. Environ Toxicol Chem 24:3092–3098.

Robertson DG. 2005. Metabonomics in toxicology: a review. Toxicol Sci 85:809–822.

Roesijadi G. 1994. Metallothionein induction as a measure of response to metal exposure in aquatic animals. Environ Health Perspect 102:91–96.

Said MR, Begley TJ, Oppenheim AV, Lauffenburger DA, Samson LD. 2004. Global network analysis of phenotypic effects: protein networks and toxicity modulation in *Saccharomyces cerevisiae*. Proc Natl Acad Sci USA 101:18006–18011.

Schmitt CJ, Hinck JE, Blazer VS, Denslow ND, Dethloff GM, Bartish TM, Coyle JJ, Tillitt DE. 2005. Environmental contaminants and biomarker responses in fish from the Rio Grande and its US tributaries: spatial and temporal trends. Sci Total Environ 350:161–193.

Sheader DL, Gensberg K, Lyons BP, Chipman K. 2004. Isolation of differentially expressed genes from contaminant exposed European flounder by suppressive, subtractive hybridization. Mar Environ Res 58:553–557.

Sheader DL, Williams TD, Lyons BP, Chipman JK. 2006. Oxidative stress response of European flounder (*Platichthys flesus*) to cadmium determined by a custom cDNA microarray. Mar Environ Res 62:33–44.

Sheahan DA, Brighty GC, Daniel M, Kirby SJ, Hurst MR, Kennedy J, Morris S, Routledge EJ, Sumpter JP, Waldock MJ. 2002. Estrogenic activity measured in a sewage treatment works treating industrial inputs containing high concentrations of alkylphenolic compounds—a case study. Environ Toxicol Chem 21:507–514.

Snape JR, Maund SJ, Pickford DB, Hutchinson TH. 2004. Ecotoxicogenomics: the challenge of integrating genomics into aquatic and terrestrial ecotoxicology. Aquat Toxicol 67:143–154.

Soetaert A, Moens LN, Van der Ven K, Van Leemput K, Naudts B, Blust R, De Coen WM. 2006. Molecular impact of propiconazole on *Daphnia magna* using a reproduction-related cDNA array. Comp Biochem Physiol C Toxicol Pharmacol 142:66–76.

Stegeman JJ, Lech JJ. 1991. Cytochrome P-450 monooxygenase systems in aquatic species: carcinogen metabolism and biomarkers for carcinogen and pollutant exposure. Environ Health Perspect 90:101–109.

Tinsley D, Wharfe J, Campbell D, Chown P, Taylor D, Upton J, Taylor C. 2004. The use of direct toxicity assessment in the assessment and control of complex effluents in the UK: a demonstration program. Ecotoxicology 13:423–436.

Tom M, Auslander M. 2005. Transcript and protein environmental biomarkers in fish—a review. Chemosphere 59:155–162.

Travis CC, Bishop WE, Clarke DP. 2003. The genomic revolution: what does it mean for human and ecological risk assessment? Ecotoxicology 12:489–495.

[USEPA] US Environmental Protection Agency. 1991. Technical guidance document for water quality-based toxics control. Washington, DC: EPA/505/2-90-001.

[USEPA] US Environmental Protection Agency. 2000. Methods for measuring the toxicity and bioaccumulation of sediment-associated contaminants with freshwater invertebrates. Washington, DC: USEPA, Office of Research and Development, EPA/600/R-99/064.

[USEPA] US Environmental Protection Agency. 2002. Short-term methods for estimating the chronic toxicity of effluents and receiving waters, 4th ed. Washington, DC: EPA-821-R-02-013.

[USEPA] US Environmental Protection Agency. 2003. Draft final guidelines for carcinogen risk assessment. Washington, DC: USEPA.

[USEPA] US Environmental Protection Agency. 2004. Potential implications of genomics for regulatory and risk assessment. Applications at EPA. Washington, DC: USEPA, EPA/100/B-04/00.

van Aggelen, G. 2005. Effect of effluent from nine pulp mills in British Columbia on underyearling rainbow trout. Report prepared by Pacific Environmental Sciences Centre, Environment Canada, North Vancouver.

van Birgelen AP, Fase KM, van der Kolk J, Poiger H, Brouwer A, Seinen W, van den Berg M. 1996. Synergistic effect of 2,2′,4,4′,5,5′-hexachlorobiphenyl and 2,3,7,8-tetrachlorodibenzo-p-dioxin on hepatic porphyrin levels in the rat. Environ Health Perspect 104:550–557.

van der Oost R, Beyer J, Vermeulen N. 2003. Fish bioaccumulation and biomarkers in environmental risk assessment: a review. Environ Toxicol Pharmacol 13:57–149.

van Straalen NM. 2003. Ecotoxicology becomes stress ecology. Environ Sci Technol 37:324A–330A.

Venturino A, Rosenbaum E, Caballero de Castro A, Anguiano OL, Gauna L, Fonovich de Schroeder T, Pechen de D'Angelo AM. 2003. Biomarkers of effect in toads and frogs. Biomarkers 8:167–186.

Viant MR. 2003. Improved methods for the acquisition and interpretation of NMR metabolomic data. Biochem Biophys Res Commun 310:943–948.

Villeneuve DL, Larkin P, Knoebl I, Miracle AL, Kahl MD, Jensen KM, Makynen EA, Durhan EJ, Carter BJ, Denslow ND, et al. 2007. A conceptual systems model to facilitate hypothesis-driven ecotoxicogenomics research on the teleost brain–pituitary–gonadal axis. Environ Sci Technol 41:321–330.

Wang DY, McKague B, Liss SN, Edwards EA. 2004. Gene expression profiles for detecting and distinguishing potential endocrine-disrupting compounds in environmental samples. Environ Sci Technol 38:6396–6406.

Wang Z, Neuburg D, Li C, Zsu L, Kim JY, Chen JC, Christiana DC. 2005. Global gene expression profiling in whole-blood samples from individuals exposed to metal fumes. Environ Health Perspect 113:233–241.

Waring JF, Anderson MG. 2005. Idiosyncratic toxicity: mechanistic insights gained from analysis of prior compounds. Curr Opin Drug Discov Devel 8:59–65.

Waters MD, Fostel JM. 2004. Toxicogenomics and systems toxicology: aims and prospects. Nat Rev Genet 5:936–948.

Weeks JM, Spurgeon DJ, Svendsen C, Hankard PK, Kammenga JE, Dallinger R, Kohler HR, Simonsen V, Scott-Fordsmand J. 2004. Critical analysis of soil invertebrate biomarkers: a field case study in Avonmouth, UK. Ecotoxicology 13:817–822.

Wei Y-D, Tepperman K, Huang M, Sartor MA, Puga A. 2004. Chromium inhibits transcription from polycyclic aromatic hydrocarbon-inducible promoters by blocking the release of histone deacetylase and preventing the binding of p300 to chromatin. J Biol Chem 279:4110–4119.

Wharfe J, Tinsley D, Crane M. 2004. Managing complex mixtures of chemicals—a forward look from the regulators' perspective. Ecotoxicology 13:485–489.

Whyte JJ, Jung RE, Schmitt CJ, Tillitt DE. 2000. Ethoxyresorufin-O-deethylase (EROD) activity in fish as a biomarker of chemical exposure. Crit Rev Toxicol 30:347–570.

Wilson JY, Cooke SR, Moore MJ, Martineau D, Mikaelian I, Metner DA, Lockhart WL, Stegeman JJ. 2005. Systemic effects of arctic pollutants in beluga whales indicated by CYP1A1 expression. Environ Health Perspect 113:1594–1599.

Winter MJ, Verweij F, Garofalo E, Ceradini S, McKenzie DJ, Williams MA, Taylor MA, Butler PJ, van der Oost R, Chipman K. 2005. Tissue levels and biomarkers of organic contaminants in feral and caged chub (*Leuciscus cephalus*) from rivers in the West Midlands, UK. Aquatic Toxicol 73:394–405.

Wirgin I, Waldman JR. 2004. Resistance to contaminants in North American fish populations. Mutat Res 552:73–100.

Wong R, Tran V, Morhenn V, Hung SP, Andersen B, Ito E, Wesley Hatfield G, Benson NR. 2004. Use of RT-PCR and DNA microarrays to characterize RNA recovered by noninvasive tape harvesting of normal and inflamed skin. J Invest Dermatol 123:159–167.

6 Application of Genomic Technologies to Ecological Risk Assessment at Remediation and Restoration Sites

*Ann L Miracle, Clive W Evans,
Elizabeth A Ferguson, Bruce Greenberg,
Peter Kille, Anton R Schaeffner, Mark Sprenger,
Ronny van Aerle, and Donald J Versteeg*

CONTENTS

6.1	Background	124
6.2	Regulatory Framework	125
6.3	Remediation Technologies	125
6.4	Application of Genomics in ERAs for Remedial Activities	130
	6.4.1 Overview	130
	6.4.2 Specific Approaches	131
	6.4.2.1 Toxicity Reduction Evaluation and Direct Toxicity Assessment	134
6.5	Research Issues, Challenges, and Needs	138
	6.5.1 Choice of Organisms and Populations	138
	6.5.2 Chronic Exposure to Complex Mixtures of Chemicals	139
	6.5.3 Establishment of Assessment Endpoints	140
	6.5.4 Surrogate Species	142
	6.5.5 Noninvasive and Nondestructive Sampling	143
	6.5.6 Challenges for Interpretation	143
	6.5.7 Resources	144
	6.5.8 Risk Communication	145
6.6	Conclusion	145
References		146

6.1 BACKGROUND

The interaction between mankind and the environment has led to the need for a series of programs to reverse or remediate environmental damage. While the building of canals, roads, and buildings; the manufacture of chemicals and products to support economies worldwide; and the mining of natural resources have benefited mankind, these activities have also caused serious, potentially long-term impacts on numerous natural environments. For example, these impacts have included contamination of surface and ground water, soils, and sediments; stripping of surface soil layers; and generation of hazardous waste sites. Programs to address these issues include remediation programs targeted at hazardous waste sites and toxicity identification and reduction for toxic effluents. This workshop, in general, and this chapter, in particular, focus on the application of genomic technologies to contaminants in the environment. Thus, this chapter is primarily concerned with application of genomics to sites requiring remediation due to the presence of contaminants.

The process of addressing sites that require remediation begins with identifying the potential site; determining the chemicals, substances, and other factors (e.g., stripping of surface soil layers) that prevent a natural and/or desirable ecosystem from existing on the site; defining the boundaries of those sites; and setting objectives for the remediation process. This type of retrospective site identification, evaluation, and remediation process has not historically relied to any significant extent on the tools and techniques of genomics—in large part because genomics tools are only now being applied to the environment. This chapter reviews the remediation and restoration or recovery process and identifies opportunities within the process where genomic technologies can be used to resolve issues.

The techniques currently available and in development in the field of molecular and mechanistic biology in transcriptomics, genotyping, proteomics, and metabolomics (described in Chapter 1 and referred to collectively as genomic technologies throughout the remainder of this chapter) provide promising new approaches that can contribute to environmental assessment, remediation, and restoration activities. The sensitivity and selectivity of the techniques, when used independently or in concert with current assessment and delineation tools, could offer new insight into contaminant exposure and define mechanisms of impact. Also, the possibility exists that mixtures of chemicals would pose less of a conundrum to the risk assessor if the effects of the mixture could be elucidated.

How genomic technologies contribute to remediation programs now and in the future is the focus of this chapter. It first reviews the remediation and restoration or recovery process and the use of genomic technologies as the remediative process itself; it then identifies opportunities within remediation or restoration where genomic technologies can be used to resolve particular issues and how the research needs can be defined to achieve these goals. This chapter also explores the current regulatory needs that can be satisfied by genomic technologies research and potential field-reliable techniques. From these needs and the current state of the science and projected technologies, applications of genomic technology data and techniques derived from genomic data are explored in response to the need for assessment, remediation, and restoration.

6.2 REGULATORY FRAMEWORK

Many countries require environmental assessment and remediation of historically released contaminants in accordance with local or national laws and international treaties. These activities usually follow prescribed guidance similar to CERCLA (Comprehensive Environmental Response, Compensation and Liability Act 1980) and RCRA (Resource Conservation and Recovery Act 1990) assessment guidance of the United States. Although no formal remediation legislation is currently in use in other countries, there are waste-specific guidelines in use by European Union (EU) consortia (for example, ERMITE for mine waters; http://www.minewater.net/ermite/) or waste- or resource-specific guidelines set by Environment Canada for environmental protection (Environment Canada 2000). For the purposes of this discussion, the guidance followed under CERCLA will be used as the framework involved in remediation.

In general, these steps include a preliminary survey or investigation of the site to determine the probability of risk to human health and the environment and the general scope of the impact if present; they follow the basics outlined in the ecological risk assessment (ERA) framework (Figure 6.1). The characterization phase includes a more comprehensive sampling of chemical and biological parameters in order to better assess the risk posed by contaminants and the spatial distribution of the contaminants. If the characterization indicates a need for remedial action because of elevated risk, legal, or other considerations, viable remediation alternatives to reduce the existing risk and/or meet other criteria are evaluated. Under most prescribed guidance for cleanup, the process ends when the risk is reduced to an acceptable level and no further assessment or remediation is required. This process, however, increasingly include the addition of planning and success of restoration activities as a metric for site closure.

Restoration—the act or process involved with *assisting* the return of a perturbed environment to an ecologically viable condition or a new equilibrium considered to be ecologically equivalent to the previous unperturbed state—is a possible goal of remediation. In cases involving long-term perturbation, parallel restoration (i.e., restoration to the current situation reflected in a neighboring, usually contiguous environment) may be a desirable goal. For clarity, recovery is defined as the return of a perturbed environment to its former condition through natural succession or via targeted intervention. Full environmental recovery, in which the environment is returned to its state prior to perturbation, is a desirable but unlikely outcome because the new ecological state or equilibrium will more than likely be different from the previously unperturbed state.

6.3 REMEDIATION TECHNOLOGIES

Once a decision has been made that a remedial action will take place, the choice of a proper remediation strategy is critical. The decision may be based on effective risk reduction, compliance with laws and regulations, effectiveness of a technology, ability to implement a technology, cost and/or cost effectiveness, and acceptance of the remediation strategy by the local government and public. Although most traditional

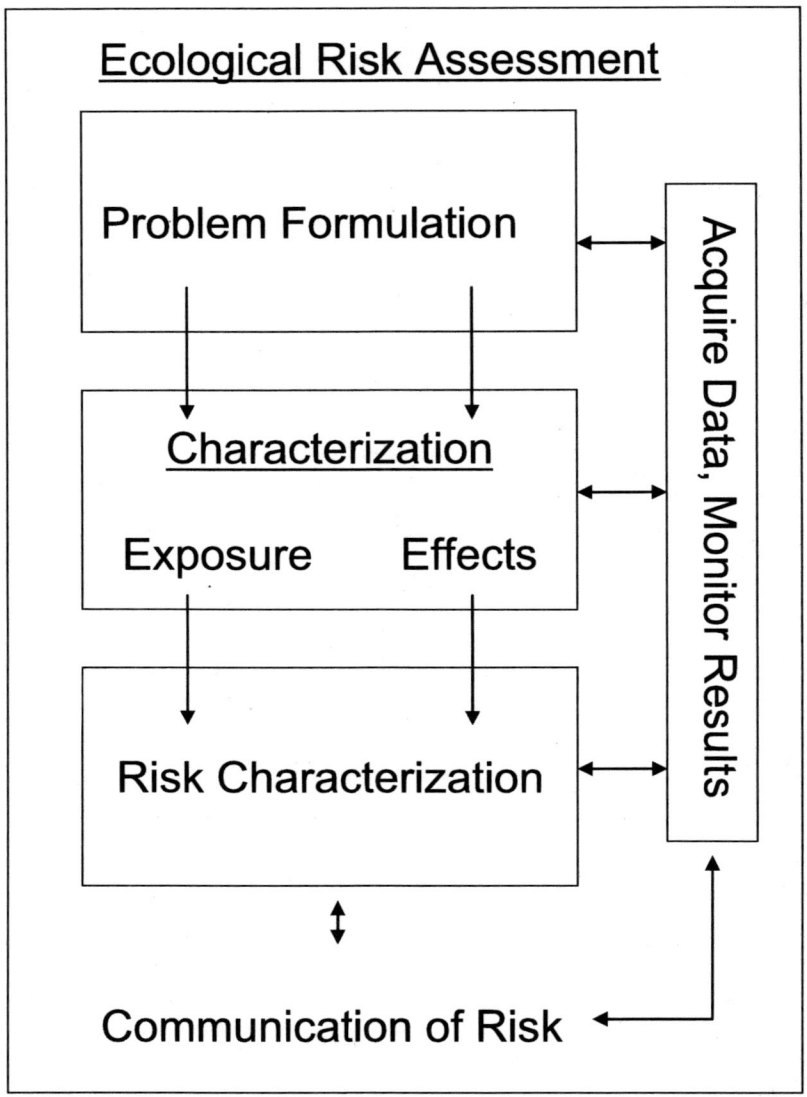

FIGURE 6.1 Ecological risk assessment (ERA) framework. (Modified from USEPA 1998.)

remediation actions have been through physical removal processes, there also are proven effective tools that include chemically and biologically based techniques. The type of environment (aquatic or terrestrial) and the nature of contaminant (organics, metals, or mixtures) are key elements for consideration of remediative action. Regardless of the choice of remedial strategy, genomics technologies can play a role in identification of contaminant types, general remediation strategies for terrestrial and aquatic systems, and monitoring remediation effectiveness. Further, if a bioremediation is part of the overall remediation strategy, genomics can play a role in the optimizing the remedy effectiveness through the selection of the organisms to be used.

Remediation is highly dependent on the major contaminant of concern, which is often broken down into 2 broad categories: organics and inorganics. Major organic contaminants include polycyclic aromatic hydrocarbons (PAHs), petroleum hydrocarbons, chlorinated organic compounds (e.g., polychlorinated biphenyls [PCBs], dioxins, chlorinated solvents, and pesticides), as well many other industrial compounds. The major inorganic contaminants are metals and salts.

In the case of organics, there is the potential to degrade the contaminants with the ultimate goal of mineralization to carbon dioxide (Bouwer and Zehnder 1993). This opens up many strategies that include in situ chemical degradation, incineration, and bioremediation. If bioremediation is chosen, the media conditions can be modified from an oxidizing environment for PAHs, petroleum, and pesticides to a reducing environment for chlorinated compounds (Wu et al. 2002).

Traditionally, inorganic contamination is either removed (with potential reuse), isolated (i.e., capped), or stabilized on-site (e.g., made into concrete). Chemical form or speciation is critical in actual toxicity and thereby the risk posed by inorganics. The concept of using alteration in chemical speciation as a mechanism for contaminant remediation in situ can be a viable option in some settings. Utilization of biotechnologies to promote changes in contaminant mobility or chemical speciation to effect risk reduction is a promising area of evaluation. It should be noted that if biological or physical stabilization on-site is chosen, environmental monitoring may be required, as is the case for traditional isolation type remedies. In many cases the contamination will be mixtures of organics and metals. This may require a phased remediation. For instance, the organics may first be degraded in situ using an appropriate bioremediation strategy. The metals may then be stabilized and/or removed.

Many remediation systems have been developed and employed to remove pollutants (especially organics) from soils (Daugulis 2001; Mulligan et al. 2001). Both physical and biological processes can be used for remediation of terrestrial systems. Physical remediation is a commonly used technique for terrestrial systems. It ranges from removal to land fill (digging and dumping) and soil washing (inorganics and organics) to on-site incineration (organics). These processes tend to be very expensive, but fast, and may not require biological or chemical monitoring to assess success of remediation.

Bioremediation has been used extensively to remediate soils, especially those contaminated with organic chemicals. Available techniques range from passive to active. The most passive is natural attenuation and/or land farming. In both cases, one generally only tills the soil and allows the natural population of soil microbes to degrade contaminants. These processes have the advantage of in situ remediation, minimal intervention, and low cost. However, the effectiveness is generally limited to the removal of readily degradable compounds such as small and/or volatile chemicals. To improve the effectiveness of natural attenuation and land farming, bio- and nutrient augmentation can be performed. Supplements such as nitrogen and phosphorus have been applied to enhance natural microbial degradation of contaminants. For degradation of chlorinated compounds, electron donors can be supplied to promote reductive dechlorination (Wu et al. 2002). These are generally metabolic sugars such as lactate. However, even with this enhancement, remediation can be slow and incomplete.

For biological remediation to be effective, the throughput must be very high (Cunningham and Ow 1996). A route for achieving this is by increasing biomass. For this reason, phytoremediation has received considerable attention in the past decade (Cunningham and Ow 1996; Salt et al. 1998; Macek et al. 2000; Pulford and Watson 2003; Singh and Jain 2003). Plants have extensive root systems that can infiltrate large volumes of soil and assimilate contaminants over a wide area. In addition, roots can enhance microbial activity by supplying substrates and nutrients. Phytoremediation with a variety of plant species (ranging from grass to trees) has been used successfully to remediate numerous environmental contaminants, including metals, herbicides, and PAHs (Meagher 2000; Mattina et al. 2003; Pulford and Watson 2003; Singh and Jain 2003; Huang et al. 2004a).

Although using plants to remediate persistent organic contaminants presents advantages over other methods, many limitations exist for current application on a large scale (Cunningham and Ow 1996; Salt et al. 1998; Glick 2003; Huang et al. 2004b). For instance, when pollutant concentrations in the soil are high, many plants cannot produce sufficient biomass for successful remediation. Often, contaminated soils are poor in nutrients; this limits plant growth and slows the remediation process. Microbial populations are often limited in contaminated soils, and they frequently lack diversity, which can further inhibit plant growth and microbial degradation of contaminants.

Rates of in situ biological remediation of persistent organic pollutants such as PAHs usually follow second-order kinetics (White et al. 1999). Because of the exponential relationship between time and contaminant concentration in soil, it takes a long time for a single remediation process to remove these types of pollutants completely. An additional challenge is that contaminated sites usually contain complex mixtures of pollutants. It is virtually impossible to remove all of the components of complex mixtures efficiently and effectively using a single technique.

To address the problems inherent in single-process remediation, one can use plant-microbe systems (Huang et al. 2004a, 2004b). One such interaction is plants plus plant-growth-promoting rhizobacteria (PGPR) (Artursson et al. 2006). The bacteria promote plant growth by, among other things, preventing production of stress ethylene and reducing the deleterious effects of plant pathogens (Glick 1995; Garcia de Salamone et al. 2001; Jetiyanon and Kloepper 2002). This trait is provided naturally by the bacteria via aminocyclopropane carboxylic acid (ACC) deaminase. With such systems, the plants can achieve large amounts of biomass in the soil even under extreme stress conditions. The roots then provide a substrate for a variety of microbes, and several biological systems together can degrade the mixtures of contaminants.

For metals, remediation using microbes can promote biostabilization (Stearns et al. 2005). For example, wetland sediment systems can be used to promote anoxic conditions, which promote sulfur-reducing bacteria. The generation of sulfides then promotes the precipitation of metal sulfides that have low solubility and, in the case of Pb, can lead to the formation of galena, thereby effectively removing the Pb from the biological system (Skowron and Brown 1994). For instance, if trees are planted and they take up the metals, the metals will be largely unavailable to other organisms during the lifetime of the tree. When remediation is complete, the trees and stumps can be removed to another site. Metals can also be taken up by tuberous plants and removed to a facility where the metals can be isolated from the plant tissue. Additionally, some plants will translocate metals to their shoots, which can be isolated and removed to a recovery facility.

Within aquatic systems, the contamination will be in either the water column or the sediment. The remedial strategy adopted will vary greatly based on the compartment. In the case of the water column, microbes and floating macrophytes can be used. For instance, bacteria have been used that can consume oil from oil spills (Harayama et al. 1999) and macrophytes have been used successfully to lower nitrate and phosphate levels in sewage (Gijzen 2001). The main challenges for remedial strategies dealing with water column contamination involve issues concerning the hydrology of the system, including problems associated with movement of the contamination.

Biotransformation refers to the chemical alteration of a substance by a living organism. It usually involves enzymatic action and often (but not always) results in increased water solubility and reduced toxicity. Xenobiotics may not be targeted naturally by biotransforming enzymes, but in many cases there is sufficient similarity between the natural substrate and a xenobiotic for the latter to be degraded. This is the case with the fungus *Phanerochaete chrysosporium*, which causes white rot in wood through the action of a ligninase. Because of the chemical similarity between lignin in wood and PAHs, the latter can be at least partially biotransformed by the fungal enzyme. Microorganisms such as *Phanerochaete* thus offer potential as biotransforming agents in the remedial process for PAHs (Pointing 2001).

The secondary metabolism of plants is also flexible and able to act on xenobiotics as long as they may be taken up by the plant (Pichersky and Gang 2000; Messner et al. 2003). Furthermore, because many bacteria-biotransforming enzymes are carried on plasmids, the potential exists to use the transfection process to generate other biotransforming hosts, possibly more effective in specific environments. In many cases, a single microorganism may lack the necessary enzymes to complete a biotransforming process. In such situations, different microorganisms that act sequentially in the degradation process may be utilized, or a single species might be genetically engineered to contain the necessary enzymes for sequential degradation. Genomic technologies could assist in the characterization and selection of suitable biotransforming microbial species as well as in the enhancement of biotransforming processes in existing species through genetic modification.

One potential use for molecular biological and genomic techniques in remediation is the selection of microbes and plants for specific remediation needs. This would include selection of plants based on their genetic makeup or use of genetically modified organisms (GMOs). For instance, plants that are hyperaccumulators of metals have been used for remediation (Clemens et al. 2002). The genes responsible for this property are now characterized and genomics could be used to rapidly select other species with similar traits. Genetically modified organisms have also been generated for remediation, such as expression of the ACC deaminase gene (see earlier discussion). Metal tolerance has been added to plants and aided in phytoremediation (An et al. 2006). The drawbacks of GMOs include concerns about release into the environment and the observation that they often do not work as well in the field as they do in the lab (Bates et al. 2005).

6.4 APPLICATION OF GENOMICS IN ERAS FOR REMEDIAL ACTIVITIES

6.4.1 Overview

A practical need within the field of ecological retrospective risk assessment is conducting necessary assessment with equal or greater specificity or accuracy more quickly or more cost effectively, or both. This goal of conducting assessment better, cheaper, and faster drives the techniques used to conduct any individual assessment. Genomic technologies show promise from this perspective (see Chapter 1 for discussion of these technologies). However, in order to meet their potential, genomic technologies will need to fill a number of data needs and overcome a number of challenges.

Diagnostic measures assist ecological assessment in that they provide a direct causal link between contaminants and risk or effects. Establishment of causal links has several important ramifications; for example, there may be implications regarding the legal authority to take an action as well as who will pay for the action. Diagnostic measures may also focus the remediation on those contaminants that pose the greatest risk, thereby reducing the cost of the remediation, by directing remediation strategies away from technology; this would be ineffective because it would be directed at the wrong contaminant. The availability of diagnostic measures would allow actions taken to be focused upon the contaminants directly responsible for risk and potentially result in significant cost savings in terms of remedy cost and in documentation of the remedy success.

Rapid, cost-effective tools to assess effects and/or exposure (bioavailability) would be a great asset to the remediation process. Many of the tests currently used in site evaluation are time consuming and costly. Tests such as solid phase toxicity tests can have time frames from sample collection to data evaluation of weeks to months. Assays, which yield results in short time frames, allow greater flexibility in sampling design, and this can result in a better understanding of the site and, ultimately, lower overall site costs. Tests, which are in and of themselves inexpensive, add to the attractiveness of the use of these methodologies.

As noted in Suter (2000), an important criterion for the selection of individual assessment endpoints for an ERA, regardless of the phase of the site investigation, is knowledge of the receptors' toxicological sensitivity to the contaminants being evaluated. For many ecological receptors there is no specific information on contaminant susceptibility; therefore, susceptibility is assumed. The use of genomic techniques could provide insight into the endpoint selection by identifying candidate assessment endpoints likely to be toxicologically susceptible to the site contaminants.

Genomics also can provide new tools for investigation of alternate assessment endpoints (e.g., soil community of ecosystem function such as nitrogen cycling)—a point discussed in more detail in the following section. A number of ecologically relevant populations and or communities have not been traditionally evaluated in ecological risk assessment. These include microbial communities and soil communities and/or environmental functions, as well as herptile receptors. One reason for this omission is a lack of viable measurement techniques and the expense of creating necessary extensive data sets. Genomics shows great promise in providing the means to include and evaluate these groups (references to techniques provided in Chapter 1).

Alternate tools to complement existing tools such as long-term toxicity tests may be another important role for genomics. As noted previously, standard toxicity testing is currently used for the evaluation of risks. Genomic techniques may provide alternatives to standard toxicity tests and/or a validated bioassay, which can be used in conjunction with toxicity testing to provide information on existing effects, residual effects, or reduced exposure that would be anticipated to relate to reduced effects. Mixtures of contaminants currently present a challenge to the development of causal relationships and targeted remediation. Use of genomic technologies to hone in on the specific mechanism of chemical impact may significantly improve the success of remedial actions as far as achieving remedial goals and also improve the restoration process.

Genomic tools that can be used to evaluate the performance or success of a remediation or corrective action are similarly needed. A large challenge for remediation evaluation is the determination of completion or, more importantly, success. Success at hazardous waste sites has traditionally been measured in terms of reduction in contaminant bulk concentration and attaining the lowest observable effect levels for standard toxicity model organisms. Some ongoing efforts to develop remediation technologies rely on reduction in bioavailability rather than reduction in absolute concentration. For these types of remedy alternatives, different measures of remediation performance must be developed and utilized. Conceptually, this is the same as an evaluation of sediment effect levels through the application of equilibrium partitioning. Genomic technologies show promise as tools for the evaluation of meeting remediation performance goals independently of bulk chemistry measurements.

Throughout the investigation, remediation, and restoration process, genomic technologies may offer appropriate and valuable tools. The methodologies employed could extend across different levels of biological organization, from the individual species to populations and communities. Some tools and methods are currently available, while others are considered future goals. Genomics information in the form of molecular or metabolic endpoints could improve the process of retrospective risk assessment used throughout the remedial process by providing faster qualitative and quantitative chemical exposure analysis, as well as information on impacts through mechanism elucidation of toxicity or direct measures of deleterious physiological reactions (Figure 6.2). This would provide more accurate biological representation of toxicity with the ability to perform rapid assessment. Specific discussion of these technologies is conducted in the following section.

6.4.2 Specific Approaches

The initial assessment, in which rapid measures of exposure are required to provide contaminant delineation, may benefit from future genomics techniques that are simple "dipstick" indicators that provide a relatively rapid and cost effective measurement tool. An example of a genomic technology that could be used might be a bio-electric switch that triggers a transcript- or protein-level response in the presence of a contaminant. This technology would indicate presence but not concentration of

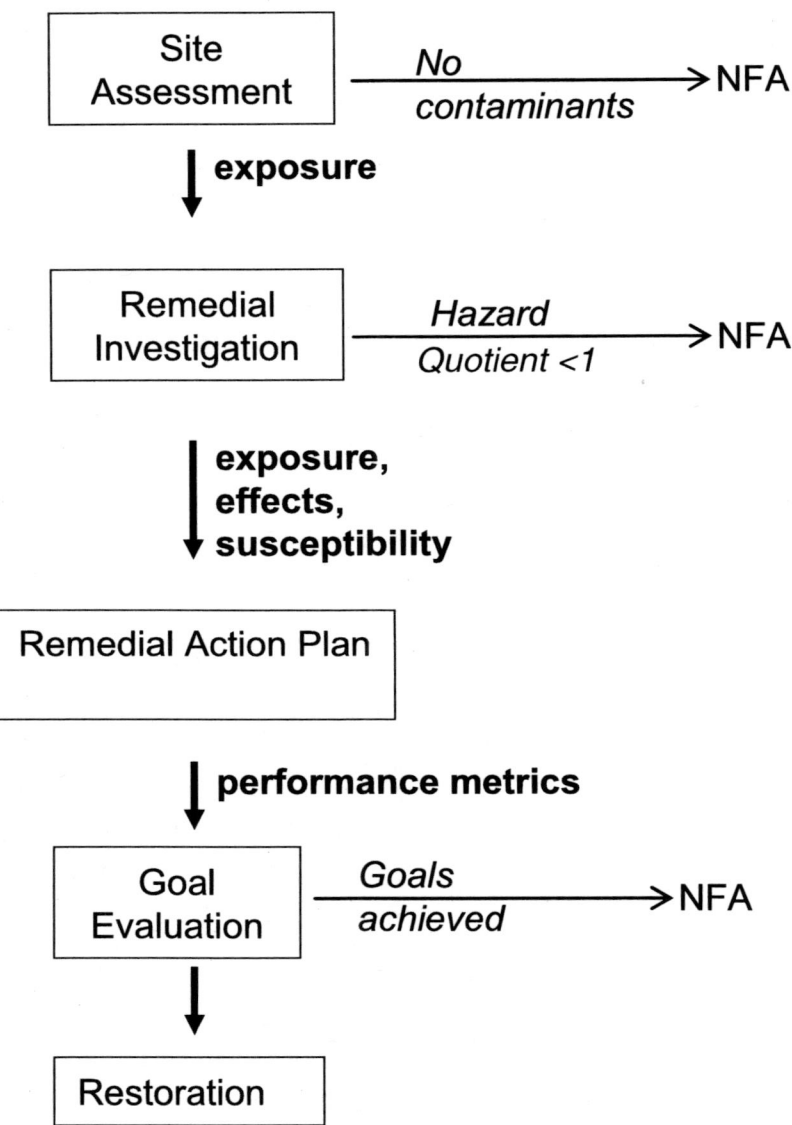

FIGURE 6.2 Areas (bolded) within the characterization of ecological risk assessment in which genomic technologies can provide clarity within the remediation framework (boxes). NFA = no further action

the contaminant by a change in on/off configuration. Antibody arrays containing hundreds of antibodies specific for proteins important in various cellular processes are a promising new tool for the analysis of protein changes in cells responding to different stimuli (e.g., Kopf et al. 2005). A dipstick version similar to the lateral flow chromatography commercial pregnancy test with a genomics-based technology complement may provide a similar simple indicator of contaminant presence.

The comprehensive sampling and analysis phase for chemical and biological assessments has multiple uses for genomic techniques and technologies for exposure assessment and impact analysis. Genomic tools to assess exposure may provide a sensitive indication that a contaminant of concern is present, the extent of the exposure probability, and, more importantly, biological availability to indigenous species. Currently, organisms present at a given site are typically not compatible with the application of existing transcriptomic or proteomic approaches to assessing exposure, given the focus of these technologies on laboratory, model species. Future advances may improve the database to include these organisms. Metabolomic profiles that represent metabolites produced in response to an exposure may provide a more immediate answer to the issue of exposure. Sampling for this type of analysis (plasma, scat, etc.) is generally nonlethal and should be less species specific, thus providing more advantages in the short term compared with the other genomic technologies.

Genomic tools used at this stage can also provide better linkages on the probable effects of a toxicant and determine mechanism or mode of action of the contaminant in the organism. Many different types of genomic analyses can be employed based upon the particular media, species of concern, and types of contaminants involved. As an example, molecular or metabolite fingerprints that can characterize a mode of action within a complex environmental mixture (Oberemm et al. 2005) provide great potential for insight into remediation needs. Use of genomic indicators may result in better estimates of an allowable concentration of the chemical in the environment, thus obviating the need for unnecessary remedial actions.

The monitoring of environmental restoration using genomic technologies could utilize laboratory- or field-based approaches. In the former, one or more model systems (based on cell lines or model organisms) would be exposed to a component of the perturbed environment (e.g., soil or pond water) and the extent of the response measured using one or more of a suite of available genomic, proteomic, or metabolite procedures. Progressive restoration of the perturbed environment would be reflected in a relative improvement with time in the extent of the measured response. Field-based approaches would involve sampling representative or targeted species from within the environment and utilizing similar technological procedures to assess the response levels to specific pollutants. Nonlethal sampling and analysis may be required for at-risk or otherwise endangered species. One possible approach to dealing with such species could involve the application of genomic technologies to assess measurable changes in blood parameters, including the induction of specific genes in blood cells.

Another application of genomics technologies to the investigation process might be in assessing changes in community structure of microbes. Species diversity is sometimes used as an index of monitoring ecosystem health (Metfies and Medlin 2005), and changes can occur rapidly in microbial and fungal communities subjected to soil, sediment, or ground water remediation (Hinojosa et al. 2005). For example, groundwater remediation is a fertile area to apply genomics tools in multiple regards including: (a) as a basis for identifying or developing species or adapted communities of organisms capable of degrading (Nishino et al. 1992) and solubilizing via biosurfactant production (Shin et al. 2006) contaminants in subsurface soils and groundwater, and (b) use of gene arrays with microbial communities to understand the progress of remediation and community shifts caused by exposure to toxic compounds.

Microbial community arrays have been developed containing most of the known genes and pathways involved in the biodegradation of organic compounds and in metal tolerance (Rhee et al. 2004). These arrays provide information on the contaminant present via induction of metabolic pathways (Wu et al. 2001), relative level of contamination via microbial community diversity (Fields et al. 2005), and detection and identification of microorganisms via reverse sample genome probing (Voordouw et al. 1991). These and other emerging tools allow for the assessment of microbial communities at remediation and restoration sites. With appropriate controls and spatial and temporally balanced sampling designs, the progress of xenobiotic attenuation and degradation can be monitored as well as the impact of the remediation process on community structure and function.

The risk question will dictate which type of genomics technology method or measure should be conducted. Table 6.1 provides an illustration of how different genomics technologies can be used for different scenarios, with their advantages. Currently, logistical details may dictate which of these methods is more feasible, but the tools for investigation into exposure and effect could be available for either situation. As mentioned earlier, metabolite profiles indicative of an exposure event do not need species-specific gene sequence data to be an effective diagnostic tool. Therefore, species choice is less of a limiting factor for sampling multiple phyla in the field. In the event that species collections are not possible, ex situ assays with the contaminated media in a controlled environment can be used with relevant model organisms to better define a specific mode of action using transcriptomic and/or proteomic measures that have linkages to probable effects, and measures of susceptibility. As more species-specific sequence information becomes available, the limitations of these latter 2 technologies will decrease for nonmodel organisms, and more options in screening for threshold values for the chemicals of concern will be provided.

Genomic technologies can also play a role in the formulation and attainment of remedial goals. The same genomic measures used to characterize risk (see Chapters 2 and 4) could often be employed to monitor the progress of attainment of remedial goals and be measures of performance that levels of contaminants have reached acceptable risk levels in the species or surrogate species of concern. To illustrate, it is conceivable that measures of particular gene transcripts, metabolites, and/or proteins can be used to indicate a reproductive impact in an organism when exposed to a chemical with a specific mode of action. Concurrent use of the genomics techniques with determination of chemical concentrations would provide validation for the remedial performance measurements or serve as the actual metrics. As an example of a specific remediative process, the next section details where genomics technologies could play a key role in the evaluation, investigation, action plan, and establishment of endpoints for direct toxicity assessment of effluents.

6.4.2.1 Toxicity Reduction Evaluation and Direct Toxicity Assessment

Effluents from manufacturing plants and municipal and privately owned wastewater treatment plants are discharged into the aquatic environment in all industrialized

TABLE 6.1
Selected Examples of the Application of Genomic Technologies to Remediation[a]

Classification	Form	Examples	Genomic Application	Benefits
Physical	Natural attenuation	Runoff; dilution; decay	Indirect	Process monitoring
	Degradation; incineration	Chemical processing of chlorinated compounds	Indirect	Process monitoring
	Stabilization	Chemical trapping metals	Indirect	Process monitoring
	Removal	Trucking and dumping	Indirect	Process monitoring
Biological	Natural attenuation	Microbial degradation	Metabolomics	Sensitive, cost-effective, and reliable monitoring of the process
	Bioaccumulation	Assimilation into an organism	qPCR for heavy metals	Accurate assessment via genomic biosensor
	Biotransformation; biodegradation	Solubilization of soil contaminants	Microarray technology	Characterization and selection of active microbial species
	Bioaugmentation	Plant growth promoting rhizobacteria	Microarray technology; proteomics	Process monitoring
	Phytoremediation	Removal of organics and metals by plants	Microarray technology; proteomics	Process monitoring
	GMOs	Genetically modified Pseudomonas	Genome manipulation	Generation and characterization of remedial GMOs

[a] Remediation process is given as the form of remediation, with examples of a scenario where the form occurs. Physical classification may use genomic technologies in an indirect manner by process monitoring for exposure to contaminants. Genomic application is identified as a direct tool for biological classification with benefits from using the corresponding genomic technology.

countries. Compounds not removed by the treatment process have the potential to cause effects in the receiving streams and their sediments, leading to environmental degradation. Because stopping the flow of effluent into the environment is not always a viable solution, methods are needed to determine the compound or compounds in the effluent causing adverse effects and to reduce or eliminate these compounds from the effluent via treatment and source reduction.

To identify the compounds causing effects in the effluent, regulatory agencies have developed procedures referred to as toxicity identification and evaluation (TIE; USEPA) and the direct toxicity assessment (DTA; UK–EU) (Tinsley et al. 2004). These programs are managed under the National Pollution Discharge and Elimination System (NPDES) in the United States and the IPPC (Integrated Pollution Prevention and Control) Directive in Europe. These procedures have been extended to sediments to help identify and remediate impacts in this compartment (Burton et al. 2004; Kwok et al. 2005).

Direct biological monitoring of effluents provides information on biological quality of the effluent, accounting for effects of mixtures of chemicals, bioavailability, pH, and other water quality variables. Further, effluent toxicity testing measures the response to all components of the effluent whether or not they have been previously identified and can be measured by chemical analysis.

In the United States, effluents are assessed in acute and chronic toxicity tests with fish, algae, and invertebrates (USEPA 2002a, 2002b). If a biological response occurs at a relevant concentration or dilution in the effluent and the factors responsible for toxicity are not known, the effluent is evaluated to determine the causative factors. This evaluation process is referred to as the TIE. The TIE typically involves a series of physical and chemical treatment processes conducted in the laboratory in an attempt to remove or identify the responsible agent causing toxicity in the effluent. The physicochemical process that removes or reduces toxicity of the sample provides information on the potential identity of the class of chemicals contributing to the toxic events. Ultimately, these processes can be used to remove toxicity from the effluent by various treatments (Ma et al. 2005).

This initial step of the TIE process is followed with chemical identification using analytical chemistry, and final confirmation steps can be achieved through procedures such as identification of individual toxicants in the complex mixture and re-evaluating the overall toxicity of the sample. Ultimately, the potential exists to improve the waste treatment process at the plant or go up the pipe using chemical or biological tests to identify and remediate the toxicity source. Genomic tools can impact the TIE process in a variety of ways, including identification of the toxic mode or mechanism of action of causative toxicants, provision of sensitive tools to track toxicity, and increasing the speed of assessing the biological response—all factors that ultimately aid in the extrapolation of data to other aquatic species.

6.4.2.1.1 Compound Identification

The identification of "classes" of compounds results from the fact that similarly acting chemicals cause the same types of gene expression changes. In fact, the chemicals may not be structurally related at all but may elicit the same biological response. For example, nonylphenol and estradiol have little in common structurally, but both elicit estrogenic effects in vertebrates and can result in similar gene expression, protein expression, and metabolite profile changes. Hence, detection of a specific mode of action (e.g., endocrine) or other chemical signal (e.g., metallothionein or CYP1A induction) could be used with other TIE procedures to identify the compounds causing effects in the effluent. In this manner genomics would increase the specificity of the TIE process.

6.4.2.1.2 Sensitivity

Genomic responses occur both as adaptive and direct toxicity responses to compounds. Typically, in a TIE bioassay, the biological endpoints used to assess effects are survival (acute and short-term chronic tests) and/or growth (short-term chronic tests). If the effluent contains compounds that impact reproduction or other vital biological processes (e.g., migratory behavior), the effects will not be detected by existing TIE tests. Genomic technologies have the potential to detect compensatory mechanisms at concentrations well below toxic levels and to determine adverse effects in pathways not detected in traditional assays (see Chapter 5, Figure 5.2). In both cases, genomics would increase the sensitivity of the TIE process.

6.4.2.1.3 Rapid Response

Gene expression changes can occur rapidly following exposure to exogenous chemicals, with effects observed within minutes or hours of exposure (Moggs et al. 2004). Genomic tools could be used for screening for the presence of specific contaminants (e.g., endocrine-active chemicals [EACs]) in effluents or water bodies. Gene expression profiles have already successfully been obtained for different species of fish exposed to single chemicals, including certain EACs: rainbow trout (carbon tetrachloride and pyrene; Krasnov et al. 2005), largemouth bass (estradiol, nonylphenol, and p,p'-DDE; Larkin et al. 2002), sheepshead minnow (estradiol, ethinyl estradiol, diethylstilbestrol, and methoxychlor; Larkin et al. 2003), zebrafish (ethinyl estradiol; Santos et al. 2006), and plaice (ethinyl estradiol; Brown et al. 2004).

These initial experiments using microarray technologies suggest that specific chemicals may have different profiles of gene expression, potentially making it possible to identify, for example, different mode of action (MOA) classes of EACs (e.g., estrogenic compounds; Larkin et al. 2003). It is not known how gene expression profiles will change after exposure of fish to mixtures of chemicals, but this approach can potentially be informative by looking at the different pathways that might be affected by exposure. However, much more data are needed to "anchor" these gene expression profiles to phenotypic responses and to be able to predict what the (adverse) effects will be (or are). Similarly, proteomic and metabolomic profiles could be used to unravel the biological MOAs or activities present in effluents and provide tools for the rapid assessment of toxicity (Shrader et al. 2003; Stentiford et al. 2005; Viant et al. 2005).

6.4.2.1.4 Extrapolation to Other Species

Currently, the TIE process is performed, at most, with algae, 1 or 2 invertebrates, and a fish species. It is assumed that protection of these species provides some level of protection for other species exposed to the effluent. This same assumption holds true for ERAs in general. Typically, toxicity data are generated on relatively few species and extrapolated using safety factors to concentrations expected to be protective of most or all species (and, by extension, the ecosystem in general). Through genomics, we can hope to be able to better understand susceptibilities of different species and the ability of different MOAs to affect them adversely.

6.5 RESEARCH ISSUES, CHALLENGES, AND NEEDS

The potential application of genomics technologies to risk assessment and monitoring in remediation and restoration faces significant challenges that do not arise when using genomics to assess specific chemicals (e.g., Chapters 2 through 4). In particular, the following points are at issue and inherent to diagnostic assessment such as remediation and restoration:

- There is no single organism or community that is ubiquitously present at different sites that allows a general assessment.
- Any frequently occurring indigenous organism that could be exploited will have an inherent genetic variability that might influence the data.
- There will be site-specific parameters, such as pH, climate, season, soil composition, and compactness, that will influence molecular changes.
- Chemicals will almost always occur as complex mixtures and will be differentially bioavailable at different sites.
- The effects on indigenous organisms will be chronic rather than acute and have the potential to be adaptive responses.

These are just a sampling of some of the key factors that represent challenges and areas where attention to scientific validation will be critical for the future use of genomic technologies for restoration and remediation. In order to start addressing how these issues can be managed, the following discussion focuses on some essential choices that face the research community in being able to use genomics technologies for application in retrospective risk assessment, remediation, and restoration.

6.5.1 Choice of Organisms and Populations

Specific remediation site characteristics may necessitate different approaches with respect to the organisms and tools that are to be used. First, organisms that are native to the remedial site could be sampled and analyzed (in situ analysis). Alternatively, effluents, extracts, or soil samples could be taken from the site and tested under defined laboratory conditions using model organisms like *Caenorhabditis elegans*, *Arabidopsis thaliana*, or *Danio rerio* (ex situ analysis) for which well-defined genomic tools exist.

In situ analyses using native species would conceptually result in the most relevant ecological data for a given remediation site. However, the interpretation of measured data may be more difficult to relate to processes and may introduce a degree of uncertainty because limited genetic or biochemical information is typically available on organisms other than those used as laboratory models. Therefore, any endpoints to be measured would have to fulfill even stronger requirements for robustness and generality.

Another problem related to the use of native organisms and populations is the transferability and general applicability of an assessment. The choice of broadly ubiquitous organisms or ecological functions is a sine qua non for developing assays that could be the basis for a legislative implementation. Candidates might be earthworms (in Europe), insects, or microbial communities, for example. Widely distributed plants with possible in situ assessment potential are less readily available, but some, such as duckweed (*Lemna* spp.) and naturally occurring thale cress (widely distributed in Europe, for example, and now occurring naturally in many parts of the northern hemisphere), may be useful in particular circumstances (such as contaminated lakes or soils). The lack of genetic and molecular information for these situations can be bridged if genome or cDNA sequences (expressed sequence tags [ESTs]) became available for several taxa indigenous to many sites in an overlapping fashion. In addition, due to considerable homologies across species, genomic tools could also be used for analyses of related species. Importantly, such a program would have to include an analysis of the genetic heterogeneity of the chosen taxa and of its consequences for the specific assessment endpoint to be measured (see later discussion). Overall, this would be a mid- to long-term goal (i.e., more than 5 years).

Ex situ analyses do have 2 definite advantages: (a) measurements will be less prone to variability on the field sites, and (b) well-defined model organisms with complete genome information and most completely annotated molecular pathways can be exploited. Using such an approach, microorganisms, nematodes, vertebrates (fish, mice), and plants could be included in a comprehensive assessment, which may be also regarded as representing different trophic levels (also, see following discussion). Due to the established tools for these organisms, the development of targeted microarrays would be rather straightforward. Similar considerations apply to proteomic or metabolomic approaches based on these model organisms.

Extrapolation to confamilial species, or even species within the same order, may serve as a useful surrogate where threatened or endangered species are an issue. At the very least, a focus on particular biological pathways could be validated in the receptor species of concern. Nevertheless, ex situ analyses may tend to overestimate or underestimate potential hazards in the environment (Wharfe et al. 2004). Therefore, at least in selected cases, correlation with effects at a remedial site has to be investigated. Standardized sampling and handling of material on-site will be of prime importance. However, protocols can be developed on the basis of current procedures for chemical analyses.

6.5.2 CHRONIC EXPOSURE TO COMPLEX MIXTURES OF CHEMICALS

Besides the choice of organisms and sampling procedures, there is an enormous gap in knowledge concerning how complex chemical mixtures will influence pathways in single organisms (as determined by genomic technologies) and how those effects in single organisms will in turn impact overall ecosystem structure and function. Thus, there is a need for a stepwise approach, starting from individual organisms under defined growth conditions challenged by a single chemical (in different doses) to more complex mixtures and possibly to various (native) organisms as an ultimate step. The obvious benefits of such an ultimate step need to be grounded in the investment of time and financial resources.

Many studies have addressed the toxicity of mixtures of chemicals (for reviews, see Wharfe et al. 2004; Jonker et al. 2005; Stierum et al. 2005). Most work has focused on whole-animal apical endpoints as opposed to well-defined biochemical responses. However, Brian et al. (2005) found that mixtures of 5 estrogenic compounds had additive effects on vitellogenin induction in fish when tested at environmentally relevant concentrations. Overall, however, there is a lack of information on the additive, synergistic, or even antagonistic effects of complex mixtures of chemicals when assessed at the molecular level. Thus, at least several mixtures known to occur coincidently at remedial sites would have to be investigated to extract robust genomic signatures. Theoretically, there is no limit to new combinations and chemical substances. However, a systematic and focused approach may also reveal rules of interactive relations that allow extrapolations. Such extrapolations may be also based on combinatorial experimental designs (Shasha et al. 2001). Significant research will be necessary to determine to what extent these approaches will suffice to provide a foundation for legislative directives.

In addition to the coincident occurrence of chemicals, almost any exposure at remedial sites will be chronic. This contrasts with many current genomic studies that look at disturbances within hours or, at most, within several days of exposure. Significantly, Ellinger-Ziegelbauer et al. (2004) showed a continued deregulation of genes 14 days after exposure to genotoxic carcinogens. During longer periods there may be also a substantial adaptation and the newly established homeostasis may be less drastically deviating from an undisturbed state (Villeneuve et al. 2007).

6.5.3 Establishment of Assessment Endpoints

Several studies using animals, plants, or microorganisms have shown that molecular signatures can be related to the structure, known toxicity, time, and even route of administration of chemicals (Ekman et al. 2003; Amin et al. 2004; Glombitza et al. 2004; Kier et al. 2004; Merrick and Bruno 2004; Steiner et al. 2004; Butcher and Schreiber 2005). A few researchers have even tackled the question of defined mixtures, at least of low complexity (see references in preceding sentence). These results indicate that molecular signatures can be used to discriminate and detect effects of chemical exposure. Assessment endpoints can be signatures based on transcripts, metabolites, or proteins as discussed in Chapter 2 and depicted in Table 6.2. Any general requirement for signatures such as specificity, low variability, or dose dependency will be more or less applicable to risk assessment of remedial sites as well.

However, a number of specific issues related to remediation and/or restoration require additional attention. The first steps should be conducted only with already established model organisms or species for which at least partial genomic information, such as ESTs or metabolic profiles, is available. Depending on the outcome of these studies and parallel genomic assessment of indigenous species, it may be possible to define orthologous endpoints and signatures for indigenous species.

The following issues have to be addressed:

- *Systematic analysis of the molecular data after exposure to environmentally relevant chemicals.* This will define specific and discriminating signatures that will be the basis for further experiments targeting their robustness and generality. The data may also include surveys of microbial communities (Metfies and Medlin 2005)[1].

---

[1] Currently, 16S ribosomal DNA is almost exclusively used to classify bacterial communities as genetic markers. However, it has to be stated that this is only a rudimentary and highly simplifying approach that does not reveal and differentiate most of the microbial diversity.

TABLE 6.2
Specific Genomic Tools by Molecular Compartment that Identify Relevant Application and How It Has Been Used in Environmental Assessment[a]

Molecular Target	Scale	Application	Tools	Capability Delivery	Comment	Ref.
Genome	Single	Microbial prevalence	Sequencing/PCR	S	Comprehensive coverage; need 16S sequences	Bodelier et al. 2005; Elshahed et al. 2005
	Targeted multiplex	Occurrence of functional pathway	Microarray	S	Many pathways available; genome needed	Chung et al. 2005; Lee et al. 2006
	Global	Community analyses	16S/23S Microarray	M	Comprehensive coverage; need 16S sequences	Stin et al. 2003; Lee et al. 2006
Transcript	Single	Indicators of MOA	QPCR; HPA	S	Small number of targets identified	Funkenstein et al. 2004; Greytak et al. 2005
	Targeted multiplex	Tiered effect and exposure indicators	Transcript and bead-based microarrays	M	Only available for rodent models; genome needed	Waring et al. 2003
	Global	Signatures of exposure and/or effect	Transcript microarrays	L	Genomic models and a few sentinels available; need more genome information	Mathavan et al. 2005; Soetaert et al. 2006
Protein	Single	Indicators of MOA	ELISA; lateral flow antibody dipsticks; biochemistry	S	Small number of targets	Schmitt et al. 2005
	Targeted multiplex	Tiered effect and exposure indicators	Lateral flow antibody dipsticks; antibody arrays	M	Not currently available	
	Global	Signatures of exposure and/or effect	2D gels; MS methods	L	Genomic models and a few sentinels; need genome information	Mi et al. 2005
Metabolite	Single	Indicators of MOA	NMR; MS methods	M	Metabolite libraries needed	Pincetich et al. 2005

[a] Scale refers to a single molecular endpoint (gene, protein, metabolite), several endpoints in a multiplex fashion, or global for assessing all available targets. Capability delivery is given as a time frame for S = short (1 to 2 years), M = medium (2 to 5 years), and L = long (5 to 10 years). Comments and references indicate special conditions where the tools have been used for a particular application.

- *The extension of these studies into chemical mixtures based on known coincidences (see preceding discussion) and into the sustainability of genomic signatures during chronic exposure.* The latter is less important for ex situ analyses, which may disregard chronicity. However, signature sustainability is absolutely necessary for in situ approaches.
- *Investigations on the influence of site-specific characteristics like nutrient availability, soil type and compactness, or even weather conditions on the data.* There could be considerable variation on a given site and even greater changes when referring to different sites. Thus, an important task will be to analyze whether any changes in gene or protein expression or metabolome composition are robust under different scenarios or integrating the physical and chemical data with the molecular information.
- *Generation of MOA data sets that will allow quantitative structure activity relationship (QSAR) analyses from standard effect profiles to profiles generated from exposure of model and surrogate species to samples collected from remedial sites.* This will require substantial work and the development of novel computational methods to underpin the interpretation (Shasha et al. 2001; Semeiks et al. 2006; Wolting et al. 2006). International cooperation centered around key organisms, but with an agreed training set of chemicals, would substantially enhance the future potential of these technologies.
- *Genetic information on indigenous species.* To conduct any in situ analyses it is important to obtain molecular and genetic information on selected ubiquitous organisms. It would be necessary to conduct EST sequencing projects, as well as an assessment of the genetic heterogeneity among individuals on a single site and on different sites, which may be another source of variability in their responses to chemical exposure.

6.5.4 Surrogate Species

The ex situ toxicological evaluation of contaminated soils or ground water can deliver an accurate assessment of the potential effect of exposure for an endangered species. But, 2 theoretical challenges exist: the identification of an appropriate surrogate model and the establishment of molecular-based MOA data sets that allow accurate interpretation and effect determination.

Identification of predictive surrogates can hinge on the cross-species knowledge of conserved MOAs. For well established MOAs, which are conserved within vertebrate species, the substantive rodent toxicity data sets can be exploited (Ekins et al. 2005; Rusyn et al. 2006). However, for MOAs outside the rodent or primate sphere much less data are available. There is, however, a fragile consensus on lead models that may be used to represent trophic levels and specific taxa. These models represent organisms where the genomic data are most complete and not necessarily those where the ecotoxicological information is abundant. Whether ecotoxicological knowledge of organisms where genomic coverage exists should be furthered, or whether the genomic knowledge base for established ecotoxicological sentinels (e.g., fathead minnow vs. zebrafish) should be expanded is open to question.

Some communities have reached independent consensus of models for candidate taxa. A good example is the Daphnia Genomics Consortium, which united behind *Daphnia pulex* and established the necessary genomic data for this particular species. Technological breakthroughs are challenging the established course of action, which has been to establish ecotoxicolgical data sets for genomic models. With reduction in the costs and speed for generating genomic data, we may need to reconsider whether it is more appropriate to establish genomic tools for established sentinels. What is apparent is that concerted and coordinated international action will likely resolve these issues in a relatively short time frame (2 to 5 years).

6.5.5 NONINVASIVE AND NONDESTRUCTIVE SAMPLING

Identifying direct, sublethal effects or monitoring remedial goals within a species of concern has 2 specific difficulties: how to exploit technologies using a species with little or no genomic data (see preceding discussion), and how to be noninvasive and nondestructive in the analysis. The technological solution is to develop metabolomic endpoints using plasma, urine, fecal, or hair or feather samples. The metabolome is relatively consistent and is not dependent on genomic knowledge. Key examples of this already exist from both forensic and environment studies (Iguchi et al. 2006; Johnson and Lewis 2006). The key will be to expand the libraries of metabolites identified for various biofluids and tissues and to identify profiles consistent with exposure to specific groups of chemicals. A library of metabolites from a wide range of representative phyla will be necessary for this to occur.

6.5.6 CHALLENGES FOR INTERPRETATION

A fundamental challenge in the application of genomics technologies in remedial evaluation and restoration is the ability of these measures to reflect the potential for demonstrable responses in the environment. From a remediation standpoint, this type of complete linkage, from molecular level response to individual and population level, would be ideal. However, for some genomic techniques this linkage may not be a realistic goal. Nevertheless, even in these instances, the potential for genomic technologies to provide a measurement tool in conjunction with other chemical, biological, and/or physical measures may still prove valuable for the site assessment process.

There are a number of uncertainties in the application of genomic techniques. In order for these techniques to be applied in a decision-making setting there must be knowledge of the potential for false positives, interferences, and (perhaps more importantly) false negatives. The potential for false negatives places the site manager, who is typically required to ensure remedy protectiveness, in a position of unacceptable uncertainty. Knowledge of the specificity between contaminant and test response is therefore a critical need.

This issue is of particular concern in the evaluation of mixtures. Reproducibility is a critical element in the defensibility of remedial actions and the ability to draw conclusions of remedial success. Conflicting data are typically interpreted in the least favorable way relative to the risk estimate (the conclusion is that greater risk exists) or remedial success (the conclusion is that success was not obtained). Low relative reproducibility will render any measure less useful. Knowledge of the degree or

TABLE 6.3
Considerations for Effective Use of Genomics Technologies in the Retrospective Risk Assessment Process

Stage of Environmental Evaluation Process	Estimated Resource Expenditure/Stage	Activity in Process with Potential for Use of Genomics Tools	Example of Genomic Technology Use
Initial investigation or survey	$	Rapid indicator of contaminant presence	"Dipstick" test for indication of contaminant presence
Chemical and biological assessment of site	$$–$$$	Exposure and bioavailability and/or effect assessment	Transcriptional, proteomic, or metabolomic response to contaminant with links to relevant endpoint
Remediation activities	$–$$$	Performance indicators for remedial goals or optimization of selected remedial alternative	Transcriptional, proteomic, or metabolomic response to contaminant with links to relevant endpoint
Restoration activities	$–$$$	Performance indicators for restoration goals	Transcriptional, proteomic, or metabolomic response to contaminant with links to relevant endpoint

Note: $ = Low Cost/Effort; $$ = Med. Cost/Effort; $$$ = High Cost/Effort.

range of the exposure–response relationship is critical to the utilization of genomic technologies in the area of remedy selection and remediation evaluation.

6.5.7 Resources

Resources are a major challenge associated with the field application of genomic technologies. Research resources (e.g., funding) are required to develop these techniques more fully. Furthermore, to bring even a workable, fully developed technique to the point where it is an accepted and readily available tool in a regulatory setting will require field demonstration projects and/or pilot studies. The cost of these latter studies is unlikely to be met through the research community, since by this stage the basic research should be completed. Until any individual technique has demonstrated cost-effective performance, it will not be selected as a measure that would be applied to the remediation evaluation (Table 6.3).

Site managers may be in the position to support the use of new technologies in conjunction with more traditional measures, depending on agency mandates. However, the issue remains as to who can and is willing to accept the initial added cost. As the various genomic techniques come online for specific application, opportunities need to be found where site managers can act with a degree of confidence that a particular technology has a high likelihood of success and could save substantial resources in the future. Convincing site managers of the potential of genomic technologies is most likely to ensure that the field tests necessary to bring developed genomics techniques into general use are undertaken.

6.5.8 RISK COMMUNICATION

A significant challenge within all ERAs during their undertaking relates to the level of communication between site managers and other stakeholders, including the general public. This communication needs to include how the measures used relate to the evaluation being conducted and to the conclusions being drawn. The use of genomic technologies provides a significant challenge in this regard, since most stakeholders would be unfamiliar with these technologies. Failure to adequately communicate the linkage between genomic measures and what is valued by stakeholders could result not only in problems with a particular site, but also in the dismissal of the new technology by the stakeholders.

6.6 CONCLUSION

There is a significant need for research into different aspects of genomic technology applications for risk assessment at remediation and restoration field sites. Expanding efforts in environmental research to include genomic and related biological sciences will provide an understanding of exposure risk and benefits of mitigation strategies. Challenges exist with both in situ and ex situ approaches, particularly in relation to the problems of chemical mixtures and robust response signatures. Although there are currently no cost-effective, short-term genomic technology applications in retrospective risk assessment, the advantages of genomics in remediation and restoration as outlined in this chapter are a driving force toward development and proof of concept application.

The development of a genomic approach to risk assessment will require considerable financial commitment. It will be primarily the task of the academic community, national or multinational research institutions like the US National Center for Toxicogenomics (Research Triangle Park, North Carolina) or the European Joint Research Center (Ispra, Italy) and public funding agencies to conduct and to support such research. After a phase of 5 to 10 years to discover and to establish potential molecular indicators and genomic profile signatures, the knowledge has to gain acceptance within the scientific community before it can be adopted by regulatory authorities.

Intermediate to acceptance within the regulatory community, the costs for developing new molecular approaches and novel genomics technologies will need to be addressed. The industrial community, with the interest for development of cost-effective assays and methods for commercialization, will need to be involved. Even this intermediate step represents a longer term goal. Establishment of good laboratory practice (GLP) methods will be needed to standardize assays that bear up under the scrutiny applied to any test that has regulatory or legal ramifications. While some of the methods may originate in the research sector, the development of commercialized test kits for specific contaminants and specific species will go a long way toward promoting the use of genomic technologies for risk assessment practices.

Finally, it must be emphasized that the regulatory community should be included in linking genomic technologies with retrospective risk assessment. Communication of proof-of-concept studies, demonstration field studies, parallel assessments using genomic technologies, or other validation studies is vital to having the technologies accepted in the policy-making arena.

REFERENCES

Amin RP, Vickers AE, Sistare F, Thompson KL, Roman RJ, Lawton M, Kramer J, Hamadeh HK, Collins J, Grissom S, et al. 2004. Identification of putative gene based markers of renal toxicity. Environ Health Perspect 112:465–479.

An ZZ, Huang ZC, Lei M, Liao XY, Zheng YM, Chen TB. 2006. Zinc tolerance and accumulation in *Pteris vittata* L. and its potential for phytoremediation of Zn- and As-contaminated soil. Chemosphere 62:796–802.

Artursson V, Finlay RD, Jansson JK. 2006. Interactions between arbuscular mycorrhizal fungi and bacteria and their potential for stimulating plant growth. Environ Microbiol 8:1–10.

Bates SL, Zhao JZ, Roush RT, Shelton AM. 2005. Insect resistance management in GM crops: past, present and future. Nat Biotechnol 23:57–62.

Bodelier PL, Meima-Franke M, Zwart G, Laanbroek HJ. 2005. New DGGE strategies for the analyses of methanotrophic microbial communities using different combinations of existing 16S rRNA-based primers. FEMS Microbiol Ecol 52:163–174.

Bouwer EJ, Zehnder AJ. 1993. Bioremediation of organic compounds—putting microbial metabolism to work. Trends Biotechnol 11:360–367.

Brian JV, Harris CA, Scholze M, Backhaus T, Booy P, Lamoree M, Pojana G, Jonkers N, Runnalls T, Bonfà A, et al. 2005. Accurate prediction of the response of freshwater fish to a mixture of estrogenic chemicals. Environ Health Perspect 113:721–728.

Brown M, Davies IM, Moffat CF, Robinson C, Redshaw J, Craft JA. 2004. Identification of transcriptional effects of ethynyl oestradiol in male plaice (*Pleuronectes platessa*) by suppression subtractive hybridization and a nylon macroarray. Marine Environ Res 58:559–563.

Burton GA Jr, Nordstrom JF. 2004. An in situ toxicity identification evaluation method. Part II: field validation. Environ Toxicol Chem 23:2851–2855.

Butcher RA, Schreiber SL. 2005. Using genome-wide transcriptional profiling to elucidate small-molecule mechanism. Curr Opin Chem Biol 9:25–30.

Chung WH, Rhee SK, Wan XF, Bae JW, Quan ZX, Park YH. 2005. Design of long oligonucleotide probes for functional gene detection in a microbial community. Bioinformatics 21:4092–4100.

Clemens S, Palmgren MG, Kramer U. 2002. A long way ahead: understanding and engineering plant metal accumulation. Trends Plant Sci 7:309–315.

Cunningham SD, Ow DW. 1996. Promises and prospects of phytoremediation. Plant Physiol 110:715–719.

Daugulis AJ. 2001. Two-phase partitioning bioreactors: a new technology platform for destroying xenobiotics. Trends Biotechnol 19:457–462.

Ekins S, Nikolsky Y, Nikolskaya T. 2005. Techniques: application of systems biology to absorption, distribution, metabolism, excretion and toxicity. Trends Pharmacol Sci 26:202–209.

Ekman DR, Lorenz WW, Przybyla AE, Wolfe NL, Dean JFD. 2003. SAGE analysis of transcriptome in *Arabidopsis* roots exposed to 2,4,6-trinitrotoluene. Plant Physiol 133:1397–1406.

Ellinger-Ziegelbauer H, Stuart B, Wahle B, Bomann W, Ahr HJ. 2004. Characteristic expression profiles induced by genotoxic carcinogens in rat liver. Toxicol Sci 77:19–34.

Elshahed MS, Najar FZ, Aycock M, Qu C, Roe BA, Krumholz LR. 2005. Metagenomic analysis of the microbial community at Zodletone Spring (Oklahoma): insights into the genome of a member of the novel candidate division OD1. Appl Environ Microbiol 71:7598–7602.

Environment Canada. 2000. Strategic environmental assessment at Environment Canada: how to conduct environmental assessments of policy, plan and program proposals. EPSM-629.

Fields, MW, Yan T, Rhee S-K, Carroll SL, Jardine PM, Watson DB, Criddle CS, Zhou J. 2005. Impacts on microbial communities and cultivable isolates from groundwater contaminated with high levels of nitric acid-uranium waste. FEMS Microbiol Ecol 53:417–428.

Funkenstein B, Dyman A, Levavi-Sivan B, Tom M. 2004. Application of real-time PCR for quantitative determination of hepatic vitellogenin transcript levels in the striped sea bream, *Lithognathus mormyrus*. Mar Environ Res 58:659–663.

Garcia de Salamone IE, Hynes RK, and Nelson LM. 2001. Cytokinin production by plant growth promoting rhizobacteria and selected mutants. Can J Microbiol 47:404–411.

Gijzen HJ. 2001. Anaerobes, aerobes and phototrophs. A winning team for wastewater management. Water Sci Technol 44:123–132.

Glick BR. 1995. The enhancement of plant growth by free-living bacteria. Can J Microbiol 41:109–117.

Glick BR. 2003. Phytoremediation: synergistic use of plants and bacteria to clean up the environment. Biotechnol Adv 21:383–393.

Glombitza S, Dubuis P-H, Thulke O, Welzl G, Bovet L, Götz M, Affenzeller M, Geist B, Hehn A, Asnaghi C, et al. 2004. Crosstalk and differential response to abiotic and biotic stressors reflected at the transcriptional level of effector genes from secondary metabolism. Plant Mol Biol 54:817–835.

Greytak SR, Champlin D, Callard GV. 2005. Isolation and characterization of two cytochrome P450 aromatase forms in killifish (*Fundulus heteroclitus*): differential expression in fish from polluted and unpolluted environments. Aquat Toxicol 71:371–389.

Harayama S, Kishira H, Kasai Y, Shutsubo K. 1999. Petroleum biodegradation in marine environments. J Mol Microbiol Biotechnol 1:63–70.

Hinojosa, MB, Carreira JA, Garcia-Ruiz R, Dick RP. 2005. Microbial response to heavy metal-polluted soils: community analysis from phospholipid-linked fatty acids and ester-linked fatty acids extracts. J Environ Qual 34:1789–1800.

Huang XD, El-Alawi Y, Penrose DM, Glick BR, Greenberg BM. 2004a. A multiprocess phytoremediation system for removal of polycyclic aromatic hydrocarbons from contaminated soils. Environ Pollut 130:465–476.

Huang XD, El-Alawi Y, Penrose DM, Glick BR, Greenberg BM. 2004b. Responses of three grass species to creosote during phytoremediation. Environ Pollut 130:453–463.

Iguchi T, Watanabe H, Katsu Y. 2006. Application of ecotoxicogenomics for studying endocrine disruption in vertebrates and invertebrates. Environ Health Perspect 114:101–105.

Jetiyanon J, Kloepper JW. 2002. Mixtures of plant growth-promoting rhizobacteria for induction of systemic resistance against multiple plant diseases. Biol Control 24:285–291.

Johnson RD, Lewis RJ. 2006. Quantitation of atenolol, metoprolol, and propranolol in postmortem human fluid and tissue specimens via LC/APCI-MS. Forensic Sci Int 156:106–117.

Jonker, MJ, Svendsen, C, Bedaux, JJM, Bongers, M, Kammenga, JE. 2005. Significance testing of synergistic/antagonistic, dose level-dependent, or dose ratio-dependent effects in mixture dose–response analysis. Environ Toxicol Chem 24:2701–2713.

Kier LD, Neft R, Tang L, Suizu R, Cook T, Onsurez K, Tiegler K, Sakai Y, Ortiz M, Nolan T, et al. 2004. Application of microarrays with toxicologically relevant genes (tox genes) for the evaluation of chemical toxicants in Sprague Dawley rats in vivo and human hepatocytes in vitro. Mutation Res 549:101–113.

Kopf E, Shnitzer D, Zharhary D. 2005. Panorama (TM) Ab microarray cell signaling kit: a unique tool for protein expression analysis. Proteomics 5:2412–2416.

Krasnov A, Koskinen H, Rexroad C, Afanasyev S, Molsa H, Oikari A. 2005. Transcriptome responses to carbon tetrachloride and pyrene in the kidney and liver of juvenile rainbow trout (*Oncorhynchus mykiss*). Aquat Toxicol 74:70–81.

Kwok YC, Hsieh DPH, Wong PK. 2005. Toxicity identification evaluation (TIE) of pore water of contaminated marine sediments collected from Hong Kong waters. Mar Poll Bull 51:1085–1091.

Larkin P, Folmar LC, Hemmer MJ, Poston AJ, Denslow ND. 2003. Expression profiling of estrogenic compounds using a sheepshead minnow cDNA macroarray. EHP Toxicogenomics 111:29–36.

Larkin P, Sabo-Attwood T, Kelso J, Denslow ND. 2002. Gene expression analysis of largemouth bass exposed to estradiol, nonylphenol, and p,p'-DDE. Comp Biochem Physiol B Biochem Mol Biol 133:543–557.

Lee DY, Shannon K, Beaudette LA. 2005. Detection of bacterial pathogens in municipal wastewater using an oligonucleotide microarray and real-time quantitative PCR. J Microbiol Methods 65:453–467.

Ma M, Li J, Wang Z. 2005. Assessing the detoxication efficiencies of wastewater treatment processes using a battery of bioassays/biomarkers. Arch Environ Contam Toxicol 49:480–487.

Macek T, Mackova M, Kas J. 2000. Exploitation of plants for the removal of organics in environmental remediation. Biotechnol Adv 18:23–34.

Mathavan S, Lee SG, Mak A, Miller LD, Murthy KR, Govindarajan KR, Tong Y, Wu YL, Lam SH, Yang H, et al. 2005. Transcriptome analysis of zebrafish embryogenesis using microarrays. PLoS Genet 1:260–276.

Mattina MI, Lannucci-Berger W, Musante C, White JC. 2003. Concurrent plant uptake of heavy metals and persistent organic pollutants from soil. Environ Pollut 124:375–378.

Meagher RB. 2000. Phytoremediation of toxic elemental and organic pollutants. Curr Opin Plant Biol 3:153–162.

Merrick BA, Bruno ME. 2004. Genomic and proteomic profiling for biomarkers and signature profiles of toxicity. Curr Opin Mol Theor 6:600–607.

Messner B, Thulke O, Schaeffner AR. 2003. Arabidopsis glucosyltransferases with activities toward both endogenous and xenobiotic substrates. Planta 217:138–146.

Metfies K, Medlin L. 2005. Ribosomal RNA probes and microarrays: their potential use in assessing microbial biodiversity. Methods Enzymol 395:258–278.

Mi J, Orbea A, Syme N, Ahmed M, Cajaraville MP, Cristobal S. 2005. Peroxisomal proteomics, a new tool for risk assessment of peroxisome proliferating pollutants in the marine environment. Proteomics 5:3954–3965.

Moggs JG, Tinwell H, Spurway T, Chang HS, Pate I, Lim FL, Moore DJ, Soames A, Stuckey R, Currie R, et al. 2004. Phenotypic anchoring of gene expression changes during estrogen-induced uterine growth. Environ Health Perspect 112:1589–1606.

Mulligan CN, Yong RN, Gibbs BF. 2001. An evaluation of technologies for the heavy metal remediation of dredged sediments. J Hazard Mater 85:145–163.

Nishino SF, Spain JC, Belcher LA, Litchfield CD. 1992. Chlorobenzene degradation by bacteria isolated from contaminated groundwater. Appl Environ Microbiol 58:1719–1726.

Oberemm A, Onyon L, Gundert-Remy U. 2005. How can toxicogenomics inform risk assessment? Toxicol Appl Pharmacol 207:S592–598.

Pichersky E, Gang DR. 2000. Genetics and biochemistry of secondary metabolites in plants: an evolutionary perspective. Trends Plant Sci 5:439–445.

Pincetich CA, Viant MR, Hinton DE, Tjeerdema RS. 2005. Metabolic changes in Japanese medaka (*Oryzias latipes*) during embryogenesis and hypoxia as determined by in vivo 31P NMR. Comp Biochem Physiol C Toxicol Pharmacol 140:103–113.

Pointing SB. 2001. Feasibility of bioremediation by white-rot fungi. Appl Microbiol Biotechnol 57:20–33.

Pulford ID, Watson C. 2003. Phytoremediation of heavy metal-contaminated land by trees—a review. Environ Int 29:529–540.

Rhee SK, Liu X, Wu L, Chong SC, Wan X, Zhou J. 2004. Detection of genes involved in biodegradation and biotransformation in microbial communities by using 50-mer oligonucleotide microarrays. Appl Environ Microbiol 70:4303–4317.

Rusyn I, Peters JM, Cunningham ML. 2006. Modes of action and species-specific effects of di-(2-ethylhexyl)phthalate in the liver. Crit Rev Toxicol 36:459–479.

Salt DE, Smith RD, Raskin I. 1998. Phytoremediation. Annu Rev Plant Physiol Plant Mol Biol 49:643–668.

Santos MM, Micael J, Carvalho AP, Morabito R, Booy P, Massanisso P, Lamoree M, Reis-Henriques MA. 2006. Estrogens counteract the masculinizing effect of tributyltin in zebrafish. Comp Biochem Physiol C Toxicol Pharmacol 142:151–155.

Schmitt CJ, Hinck JE, Blazer VS, Denslow ND, Dethloff GM, Bartish TM, Coyle JJ, Tillitt DE. 2005. Environmental contaminants and biomarker responses in fish from the Rio Grande and its US tributaries: spatial and temporal trends. Sci Total Environ 350:161–193.

Semeiks JR, Rizki A, Bissell MJ, Mian IS. 2006. Ensemble attribute profile clustering: discovering and characterizing groups of genes with similar patterns of biological features. BMC Bioinformatics 7:147.

Shasha DE, Kouranov AY, Lejay LV, Chou MF, Coruzzi GM. 2001. Using combinatorial design to study regulation by multiple input signals. A tool for parsimony in the postgenomics era. Plant Physiol 127:1590–1594.

Shin KH, Kim KW, Ahn Y. 2006. Use of biosurfactant to remediate phenanthrene-contaminated soil by the combined solubilization-biodegradation process. J Hazard Mater 137:1831–1837.

Shrader EA, Henry TR, Greeley MS, Bradley BP. 2003. Proteomics in zebrafish exposed to endocrine disrupting chemicals. Ecotoxicology 12:485–488.

Singh OV, Jain RK. 2003. Phytoremediation of toxic aromatic pollutants from soil. Appl Microbiol Biotechnol 63:128–135.

Skowron R, Brown I. 1994. Crystal-chemistry and structures of lead-antimony sulfides. Acta Crystall 50:524–538.

Soetaert A, Moens LN, Van der Ven K, Van Leemput K, Naudts B, Blust R, De Coen WM. 2006. Molecular impact of propiconazole on *Daphnia magna* using a reproduction-related cDNA array. Comp Biochem Physiol C Toxicol Pharmacol 142:66–76.

Stearns JC, Shah S, Greenberg BM, Dixon DG, Glick BR. 2005. Tolerance of transgenic canola expressing 1-aminocyclopropane-1-carboxylic acid deaminase to growth inhibition by nickel. Plant Physiol Biochem 43:701–708.

Steiner G, Suter L, Boess F, Gasser R, de Vera MC, Albertini S, Ruepp S. 2004. Discriminating different classes of toxicants by transcript profiling. Environ Health Perspect 112:1236–1248.

Stentiford GD, Viant MR, Ward DG, Johnson PJ, Martin A, Wei WB, Cooper HJ, Lyons BP, Feist SW. 2005. Liver tumors in wild flatfish: a histopathological, proteomic, and metabolomic study. OMICS 9:281–299.

Stierum R, Heijne W, Kienhus A, van Ommen B, Groten J. 2005. Toxicogenomics concepts and applications to study hepatic effects of food additives and chemicals. Toxicol Appl Pharmacol 207:S179–S188.

Stin OC, Carnahan A, Singh R, Powell J, Furuno JP, Dorsey A, Silbergeld E, Williams HN, Morris JG. 2003. Characterization of microbial communities from coastal waters using microarrays. Environ Monitor Assess 81:327–336.

Suter G, II. 2000. Generic assessment endpoints are needed for ecological risk assessment. Risk Anal 20:173–178.

Tinsley D, Wharfe J, Campbell D, Chown P, Taylor D, Upton J, Taylor C. 2004. The use of direct toxicity assessment in the assessment and control of complex effluents in the UK: a demonstration program. Ecotoxicology 13:423–436.

USEPA. 1998. Guidelines for ecological risk assessment. EPA-630-R-95-002Fa. US Environmental Protection Agency, Washington, DC.

USEPA. 2002a. Methods for measuring the acute toxicity of effluents and receiving waters to freshwater and marine organisms. 5th ed. EPA-821-R-02-012, US Environmental Protection Agency, Washington, DC.

USEPA. 2002b. Short-term methods for estimating chronic toxicity of effluents and receiving waters to freshwater organisms. EPA-821-R-02-013, US Environmental Protection Agency, Washington, DC.

Viant MR, Bundy JG, Pincetich CA, de Ropp JS, Tjeerdema RS. 2005. NMR-derived developmental metabolic trajectories: an approach for visualizing the toxic actions of trichloroethylene during embryogenesis. Metabolomics 1:149–158.

Villeneuve D, Larkin P, Knoebl I, Miracle AL, Kahl MD, Jensen KM, Makynen EA, Durhan EJ, Denslow ND, Ankley GT. 2007. A graphical systems model to facilitate hypothesis-based ecotoxicogenomics research on the teleost brain–pituitary–gonadal axis. Environ Sci Technol 41:321–330.

Voordouw G, Voordouw JK, Karkhoff-Schweizer RR, Fedorak PM, Westlake DW. 1991. Reverse sample genome probing, a new technique for identification of bacteria in environmental samples by DNA hybridization, and its application to the identification of sulfate-reducing bacteria in oil field samples. Appl Environ Microbiol 57:3070–3078.

Waring JF, Cavet G, Jolly RA, McDowell J, Dai H, Ciurlionis R, Zhang C, Stoughton R, Lum P, Ferguson A, et al. 2003. Development of a DNA microarray for toxicology based on hepatotoxin-regulated sequences. EHP Toxicogenomics 111:53–60.

Wharfe J, Tinsley D, Crane M. 2004. Managing complex mixtures of chemicals—a forward look from the regulator's perspective. Ecotoxicology 13:485–492.

White JC, Hunter M, Nam KP, Pignatello JJ, Alexander M. 1999. Correlation between biological and physical availabilities of phenanthrene in soils and soil humin in aging experiments. Environ Toxicol Chem 18:1720–1727.

Wolting C, McGlade CJ, Tritchler D. 2006. Cluster analysis of protein array results via similarity of gene ontology annotation. BMC Bioinformatics 7:338.

Wu L, Thompson DK, Li G, Hurt RA, Tiedje JM, Zhou J. 2001. Development and evaluation of functional gene arrays for detection of selected genes in the environment. Appl Environ Microbiol 67:5780–5790.

Wu Q, Watts JE, Sowers KR, May HD. 2002. Identification of a bacterium that specifically catalyzes the reductive dechlorination of polychlorinated biphenyls with doubly flanked chlorines. Appl Environ Microbiol 68:807–812.

7 Toxicogenomics in Ecological Risk Assessments: A Prospectus

George P Daston, Ann L Miracle,
Edward J Perkins, and Gerald T Ankley

CONTENTS

References 156

The intention of this Pellston workshop was to identify opportunities by which genomics and other high-information content methods can be pragmatically applied to aspects of ecological risk assessment. While it is still early days in terms of the amount of toxicogenomic, proteomic, and metabolomic information that has been produced, it is becoming clear that these methods have the potential to improve the scientific foundation on which risk assessments are based significantly (Ankley et al. 2006).

The most widespread use of toxicogenomics thus far has been in predictive toxicology. The idea that specific modes of toxic action would produce characteristic patterns of gene (and by extension, protein or metabolite) expression in responsive tissues has been borne out by empirical research. This in turn has supported the development of short-term assays, both in vivo and in vitro, that are capable of predicting toxicity by a selective evaluation of gene expression. A number of pharmaceutical companies, as well as specialized biotechnology firms, have developed databases on the patterns of gene expression that characterize specific modes of action and are applying these to predict early in drug discovery whether drug candidates are likely to be toxic at concentrations that are close to the concentration that is efficacious (e.g., Fielden et al. 2005; Ganter et al. 2005). Those candidates can be excluded from further development, thereby increasing the number of compounds that can be advanced further into preclinical and clinical testing and decreasing the cost of subsequent toxicity testing and drug development.

These same principles of predictive toxicology can be used in a regulatory setting for prioritizing chemicals such that those with the greatest potential to produce biological effects can be identified for further evaluation. Chemical testing prioritization for the vast majority of commodity chemicals (e.g., those regulated under the Toxic Substances Control Act in the United States) is based largely, if not exclusively, on production volumes. While this makes some sense, in that production volume is correlated with exposure potential, it is weighted entirely toward the exposure side of risk. This problem has been recognized, but it has been difficult to do anything about it because our imperfect knowledge about mechanisms of toxicity has limited the available tests for toxicity to models that focus on outcome—that is, toxicity in an intact organism.

Using simple screens for a specific mode of action was first recommended as a prioritization tool by the Environmental Protection Agency's Endocrine Disrupter Screening and Testing Advisory Committee (EDSTAC 1998). Endocrine disrupter assessment represented a fundamental change in toxic chemical regulation in that it focuses on identifying agents with particular modes of action rather than solely on apical biological outcome. Because these modes are molecular in nature (steroid hormone receptor interactions, interference with steroid synthesis pathways, or hormone catabolism), it is possible to design simple in vitro methods to assess the effects of chemicals on them, such as receptor binding assays or reporter gene assays. Such assays can be configured in such a way as to be high throughput, facilitating the assessment of large numbers of chemicals or even chemical mixtures.

Although endocrine modes of action are important because they can have adverse effects on reproduction and development, the fraction of chemicals that have endocrine-active properties is not known. There are many more modes of action that are important for ecotoxicity, and as many of these as are feasible should be included in screening batteries for which the purpose is to identify high-priority agents for further testing (Chapters 2 and 3). Genomics approaches offer the possibility of assessing a large number of modes of action, either in a single system or a limited number of simple systems that are amenable to medium or high throughput. More research needs to be done to identify relevant modes of action for ecotoxicity and the patterns of gene (or protein or metabolite) expression associated with them. The success of this approach in mammalian toxicology suggests that it will be similarly successful in ecotoxicology.

The selection of model systems is always difficult in ecological risk assessment, but the problem is much less acute in the context of upstream processes such as prioritization, where the data are unlikely to be used as the primary basis for quantitative risk assessment. Therefore, it may be possible to make more rapid progress by using model organisms such as zebrafish and *Drosophila* that have already been sequenced because of their importance in developmental biology and for which microarrays and other molecular reagents are readily available.

Predictions about mechanisms of action may also be an aid to tailoring the toxicological evaluation of a particular chemical or stressor so that the maximum amount of useful information can be obtained efficiently. If a screening battery indicates that the stressor of interest has the potential to act via a particular mode of action, then the testing plan for that stressor would be designed to evaluate in depth the biological processes known to be perturbed by that mechanism. Model systems known to be susceptible to the mechanism would be preferentially selected for testing. For example, if it were determined that a particular agent had the potential to interact with retinoic acid receptors, then it would make sense to investigate reproduction and early development in depth in model vertebrate and invertebrate species.

Although predictive toxicology and screening may be the first application of genomics in ecological risk assessment, it will not be the only application. Genomics can be applied to provide a better understanding of a compound's toxicity. The areas in which genomics could be fruitfully applied include (1) providing mode of action information as support for the risk assessment, (2) better analysis of dose–response relationships, and (3) reducing uncertainty in cross-species extrapolations (Chapters 3 and 4).

Genomics information, at its best, provides a comprehensive evaluation of potential modes of action that is not possible with current experimentation that evaluates a limited number of outcomes. This integration of information (which reaches its highest form with metabolomics, but is also the case for functional genomics and proteomics) can provide even greater certainty that the toxicology testing that has been done has adequately described the hazard. Conversely, this additional information may identify additional testing that needs to be done to determine whether an identified mode of action could lead to an outcome plausibly linked to that mode (Chapters 3 and 4).

Genomics has been shown to be a sensitive technology and may be used to better define the shape of the dose-response curve (Chapter 4). Studies that demonstrate this sensitivity and the use of genomics to define the shape of the low end of the dose-response curve for estrogens have been reported in the mammalian toxicology literature (Naciff et al. 2005). This work helped address a controversy over whether low dose effects were being produced at ultralow doses of estrogens that were not predicted by studies done at exaggerated dose levels.

This approach to risk assessment has its caveats—especially that the quantitative relationship between altered gene (or protein or metabolite) expression and frank adverse effects needs to be understood. Because of the high sensitivity of these techniques, the possibility exists that effects on gene expression can be detected that are not related to overt toxicity. This may be an advantage to the technology, but brings to light the need for carefully conducted studies that phenotypically anchor changes in gene expression.

It may also be possible to use genomics information to support intelligent testing strategies (Chapter 3). If it can be shown that the gene expression profile for a chemical that is closely related to that for a chemical with a full compendium of toxicology data, it should be possible to conclude that the toxicity of the latter is indicative of the toxicity of the former. Given the enormous testing burden that is set to occur under the EU's REACH (registration, evaluation, authorization, and restriction of chemicals) legislation, strategies like these will be important in characterizing the hazard of chemicals in an efficient and ecologically protective way, while allowing limited testing resources to be applied to evaluating less well characterized materials.

One of the uncertainties in ecological risk assessment is whether the no-observed-effect concentrations (NOECs) identified in a few lab species are adequately protective for all species (Chapters 3, 4, and 5). Gene expression data may be a tractable way to survey a broader spectrum of taxa to assure that the toxicity data generated in a few species are relevant for and quantitatively similar to responses in those taxa.

Genomics may also be useful in activities that happen after predictive risk assessments, particularly in monitoring and environmental remediation (Chapters 5 and 6). In these instances, the goal in the former case is to ensure that inputs into an ecosystem are below levels that would lead to adverse population effects and, in the latter case, to clean up a contaminated site to reach a goal—usually to restore the site to conditions that are similar to those of neighboring, unaffected sites.

In the real world, virtually all environmental exposures are to mixtures of compounds. Because genomics has the power to elucidate more than one mechanism at a time, it is possible to use these approaches to evaluate whether a mixture is producing biological effects and, if so, whether the components are acting in an additive, subadditive, or synergistic way. It is also possible to determine whether a single component of a mixture is the driver for toxicity, in which case problems of remediation and monitoring can be simplified (Chapters 5 and 6). Often, environmental samples are not well characterized, and it should be possible to use patterns of gene expression to identify mechanisms that can be used to infer exposures to classes of chemicals. This would limit the extent of characterization via more expensive methods that would need to be done.

An important application of genomics will be to determine when cleanup goals are reached at remediation sites (Chapter 6). This could be when no changes are detected in gene expression that was specific to the site contaminants or when the changes are not different from those observed in organisms from a reference population.

One of the challenges with using genomics for real-world (diagnostic) purposes is that the species present in the field may not be the same as those used as model organisms in laboratory settings (Chapters 5 and 6). It will be important, therefore, to select organisms that are likely to be environmentally ubiquitous if one wishes to use them as genomic sentinels. Because of the high degree of conservation across species at the molecular level, it may be possible to use data from lab model organisms (e.g., *Caenorhabditis elegans*, *Drosophila*) as an informatics foundation for determining changes in environmentally relevant species (e.g., nematodes, earthworms, insects). This approach is particularly attractive when determining the impact of contamination on species that are threatened or endangered, as well as the adequacy of remediation. Genomics approaches applied to tissue samples that can be collected using minimally invasive techniques may offer a way to assess both a strategy for cleanup that is optimal for the endangered population or species and the adequacy of the remediation.

Another unexplored use of genomics in field settings is to use DNA microarrays to survey the biological diversity of a site. This approach has been used to determine species diversity in ocean waters and could be more widely applied. In the context of remediation, it may be possible to use community diversity as a measure of remediation success instead of the gene expression responses of a few sentinel species. Such an approach may be closer to the ideal of ecological risk assessment and risk management. It also acknowledges the fact that, in most cases, chemical contaminants are not the only stressors in a contaminated site and that chemical-specific gene expression changes may be lost in the noise.

A number of research needs were identified in the course of the workshop and are summarized here. These include:

- studies to phenotypically anchor key genomic endpoints or patterns to adverse outcomes
- studies to distinguish responses that are directly linked with toxicity from those that are adaptive
- improvement of annotation for species that are relevant to ecological risk assessment
- determination of direct and indirect effects of nonchemical environmental variables (e.g., temperature, diet, disease) on genomic responses
- development of widely accessible toxicogenomic databases
- increased bioinformatic capabilities and expertise
- identification and validation of standardized data collection, analysis, and interpretation approaches for routine use in risk assessments
- case studies (ideally for different applications as reflected in Chapters 2 through 6)
- development and commercialization of cost-effective reagents that can be used for lab and field studies

The first two of these are not unique to ecotoxicology. Investigators who use mammalian models to predict human toxicity face the same problems. There is considerable research ongoing to link gene expression to adverse outcomes, as well as to distinguish those events that are critical for toxicity versus those that are incidental or adaptive (e.g., Ganter et al. 2005). There is also research to understand how much change in gene expression is needed to elicit an adverse effect. The research strategies being taken by mammalian toxicologists can be implemented for ecotoxicology as well. Because of conservation of gene expression across species, it will be interesting to see how much of the mammalian literature can be reapplied to environmental models.

The next two research needs are especially important for ecological risk assessment. Most of the annotation of genes in nonmammalian species has been for species that have been used as models in studying basic developmental processes: zebrafish and *Drosophila*, for example. These species are not necessarily the most relevant for ecological risk assessment, and it will be important to accelerate annotation for species that have a rich toxicological database, like fathead minnow and *Daphnia*. Because of the complexity of toxicogenomics experiments and the wealth of data produced, it will be necessary to develop large, shared databases and the tools to exploit them. Databases are being developed in a number of places for specific toxicological applications. It may be that it is possible to include data from experiments that are relevant for ecological risk assessment in these databases as well. It will also be important for reagent costs to decrease if genomics tools are to be widely used and databases populated. Finally, it will be important to start to conduct research in ways relevant to predicting ecological risk and to document this research in case studies, as a demonstration to the greater scientific community of the power of the approach.

In summary, there are a large number of potential applications for high-information-content technologies such as genomics, transcriptomics, proteomics, and metabolomics in ecological risk assessment. Some of these, such as hazard prediction and mode of action identification, are becoming relatively mature and could be used as the basis for screening or as an adjunct for risk assessment. There are a few examples in the literature for other applications, such as dose–response assessment, but much more work will need to be done before they can make their way into mainstream risk assessment. The technology will need to be further developed and applications validated, a process that will likely take several years. We believe that this further development is worth the effort because these high-information-content technologies offer the potential to address problems in ecological risk assessment that had been thought to be intractable (or bound up in irresolvable uncertainty) because of their complexity.

REFERENCES

Ankley GT, Daston GP, Degitz SJ, Denslow ND, Hoke RA, Kennedy SW, Miracle AL, Perkins EJ, Snape J, Tillitt, DE, et al. 2006. Toxicogenomics in regulatory ecotoxicology. Environ Sci Technol 40:4055–4065.

[EDSTAC] Endocrine Disrupter Screening and Testing Advisory Committee. 1998. Recommendations to USEPA on endocrine disrupter screening and testing (http://www.epa.gov/scipoly/oscpendo/edspoverview/finalrpt.htm).

Fielden MR, Eynon BP, Natsoulis G, Jarnagin K, Banas D, Kolaja KL. 2005. A gene expression signature that predicts the future onset of drug-induced renal tubular toxicity. Toxicol Pathol 33:675–683.

Ganter B, Tugendreich S, Pearson CI, Ayanoglu E, Baumhueter S, Bostian KA, Brady L, Browne LJ, Calvin JT, Day GJ, et al. 2005. Development of a large-scale chemogenomics database to improve drug candidate selection and to understand mechanisms of chemical toxicity and action. J Biotechnol 119:219–244.

Naciff JM, Hess KA, Overmann GJ, Torontali SM, Carr GJ, Tiesman JP, Foertsch LM, Richardson BD, Martinez JE, Daston GP. 2005. Gene expression changes induced in the testis by transplacental exposure to high and low doses of 17-alpha-ethynyl estradiol, genistein, or bisphenol A. Toxicol Sci 86:396–416.

Glossary

Acute: within a short period (seconds, minutes, hours, or a few days) in relation to the life span of organisms.

Adaptive risk or resource management: approaches that use qualitative or quantitative measures to monitor success or failure of a management plan and iteratively adjust the management plan until success is achieved.

Adaptive (or compensatory) response: a change that allows an organism to tolerate a toxicant—for instance, detoxification of a chemical by an enzyme.

Bioaccumulation: accumulation of chemicals in the tissues of an organism.

Bioassay: in toxicology, the use of a live organism or cell culture to characterize effects of chemicals.

Bioinformatics: branch of science that provides tools for the analysis and interpretation of large-scale data sets such as those generated from transcriptomic, proteomic, or metabolomic experiments.

Biomarker: defined as a change in a biological response (ranging from molecular through cellular and physiological responses to behavioral changes), which can be related to exposure to or toxic effects of environmental chemicals. A biomarker could also be defined as any biological response to an environmental chemical at the subindividual level, measured inside an organism or in its products (urine, feces, hair, feathers, etc.).

Body burden: the total amount of a xenobiotic substance in an organism.

Chemical monitoring: measuring selected contaminants in different environmental compartments.

Ecotoxicology: the study of the harmful effects of chemical compounds on species, populations, and the natural environment.

Effluent discharge: wastewater, treated or untreated, that flows out of a treatment plant, sewer, or industrial outfall. This term generally refers to wastes discharged into surface waters.

Environmental monitoring: activities where environmental samples are taken and analyzed to generate data for regulatory or prioritization decisions, or to develop status and trends information to better manage ecological health.

Environmental risk assessment: process of identifying and evaluating the risk for adverse effects on the environment caused by exposure to a chemical substance. An environmental exposure to the chemical is predicted (or measured) and compared to a predicted (or measured) no-effect concentration to characterize risk.

Fingerprinting: unbiased approach to measuring transcripts (or proteins or metabolites) where the specific analytes are not necessarily identified. From this, fingerprints (or "signatures") indicative of specific toxic mechanisms or modes can be derived.

Genome: the DNA sequence of an organism.

Genomics: branch of science dealing with the genetics, genomes, transcriptomes, proteomes, and metabolomes of organisms. Genomics encompasses the technologies necessary to catalog the genes of an organism and their expression (as RNA, protein, or metabolites) in response to environmental perturbations.

Imposex: a condition in which male sexual characteristics, such as the development of a penis, are superimposed on female gastropods. An example is the effect of tributylin on the common dogwhelk.

Mechanism of action: mechanism of toxic action is a molecular sequence of events from absorption of an effective dose to production of a specific biological response.

Metabolomics: large-scale study of the metabolome, the totality of metabolites in an organism, tissue, or cell type.

Metallothionein: a protein that, in addition to normal homeostatic functions, binds (detoxifies) heavy metals such as cadmium and lead.

Microarray: a technology using a high-density array of nucleic acids, protein, or tissue for simultaneously examining complex biological interactions that are identified by specific location on a solid support matrix.

Mode of action: biochemical, physiological, and cellular events leading from the specific interaction of a chemical with an endogenous target molecule (here termed "the mechanism") to manifestation of a specific biological effect.

Phenotypic anchoring: relating a specific change in gene, protein, or metabolite expression to a specific outcome at a higher level of biological organization (i.e., the cell, tissue, or organism).

Profiling: genomic analyses (e.g., to delineate toxic mode or mechanisms of action) based on identification of specific transcripts, proteins, or metabolites. In the case of transcripts or proteins, this requires some level of knowledge of genome sequence for the species of interest.

Proteomics: large-scale study of the proteome, the set of all proteins of an entire organism, tissue, or cell type.

Quantitative polymerase chain reaction (QPCR): a method of determining gene expression in small samples by amplification of specific gene targets from RNA that has been copied to cDNA. The quantity of target gene amplified is generally monitored by accumulation of fluorescent products in comparison to a known standard or reference gene.

Regulatory monitoring: actions in which industry, municipalities, or specific groups such as native peoples are mandated to conduct environmental sampling for characterization and/or control purposes.

Single nucleotide polymorphisms: a single nucleotide polymorphism, or SNP (pronounced "snip"), is a DNA sequence variation occurring when a single nucleotide—adenine (A), thymine (T), cytosine (C), or guanine (G)—in the genome is altered. A variation must occur in at least 1% of the population to be considered a SNP.

Sublethal: detrimental to organisms, but below the level that directly causes death within the test period.

Toxicity test: a method used to determine the effect of a material on selected organisms under defined conditions. An environmental toxicity test usually measures either the proportion of organisms affected or the degree of effect shown after exposure to a chemical, liquid waste, receiving water, sediment, or soil.

Toxicogenomics: branch of science dealing with the global suite of responses in gene or protein expression or metabolic profile in a cell, tissue, or organism after exposure to a toxicant.

Transcriptomics: large-scale study of the transcriptome, the set of all mRNA molecules (transcripts) of an entire organism, tissue, or cell type generally measured during a particular physiological state (i.e., responses to a perturbation are compared to the pattern of genes expressed at homeostasis).

Index

A

Acute, 157
Acute assessment, 103–105
Adaptive response
  fundamentals, 157
  human health assessments, 18
  Tier III, 104–105
Adaptive risk/resource management, 157
Alzheimer's disease, 39
Amphipods, 94, 101
Animals, *see also specific species*
  current method usage, 2
  reduction, 27, 66
  tiered testing, 47, *48*, 49–50
Ankley studies, 1–9, 151–156
Annotation availability, 110–111
Antimicrobials
  application developments, 80–83
  risk assessment, 78–79
Applications
  complex mixtures, 99–109
  remediation and restoration sites, 130–137
  tiered testing, 50–53
Aquatic organisms and systems
  pesticides, 68–69
  remediation, 129, 134–136
Arabidopsis spp., 138
Arrays availability, 110–111
ArrayTrack, 18
Assays, 15–17

B

Bioaccumulation
  fundamentals, 157
  screening level tools, 101
  Tier III, 104
Bioassay, 157
Bioassay-directed fractionation, 96
Bioinformatics
  ArrayTrack, 18
  fundamentals, 67, 157
  limitations, 111
  tiered testing, 53–54
Biological challenges, 21
Biological exposures and effects measurements, 92–93, 102–103, *103*

Biomarkers
  current monitoring approaches, 94
  ecotoxicology application examples, 20
  exposure assessment, 70–71
  fundamentals, 157
  future outlook, 110
  global *vs.* focused molecular analyses, 6–7
  multiple endpoint assays, 16
  proteomics, 41–42
  reproductive, 92
  single, 38–40
Bioremediation, 127–128
Biostabilization, 128–129
Biotransformation, 129
Birds, 41
Black box fingerprinting, 51
Body burden, 157
Bradbury studies, 63–83
Brennan studies, 63–83

C

Caenorhabditis elegans
  genome data, 3
  lifestages sensitivity, 76
  organism choice, 138
  real-world genomics usage, 154
Cancer diagnostic tests, 18
Carp, *see also* Fish
  cross-species experiments, 81
  tiered testing, 47
  transcriptomics, 41
Case-specific conditions, 105–109
Causal connections, 106
CERCLA, 125
Challenges
  complex mixtures, 96–97
  remediation and restoration sites, 138–145
  screening assays, 14
Chemical Effects in Biological Systems, 112
Chemical monitoring, 157
Chicken, 26
Chipman studies, 87–113
Chronic effects, complex mixtures, 103–105
Chronic exposure, 139–140
Chub (*Leuciscus* spp.), 40, 93
Chum salmon (*Oncorhynchus* spp.), 43
Circadian cycles, 105

161

Community structure, microbes, 133–134
Comparative Toxicogenomics Database, 112
Compensatory response, 104–105, 157
Complete genomes availability, 110–111
Complex mixtures
    acute assessment, 103–105
    annotation availability, 110–111
    application, genomics, 99–109
    arrays availability, 110–111
    biological exposures and effects measurements, 92–93, 102–103, *103*
    case-specific conditions, 105–109
    challenges, 96–97
    chronic effects, 103–105
    complete genomes availability, 110–111
    database support application, 111–112
    ecological risk assessment, 19
    effects, monitoring and assessment, 92, 93–96
    examples, current approaches of monitoring, 93–94
    exposures, monitoring and assessment, 92, 93–96
    fundamentals, 88–89, *90*
    future outlook, 110–113
    genomics development, 109–110
    genomics opportunities, 97–98
    impacted functions, 107
    individual sensitivity distribution, 108–109
    mechanism of action, 106
    mode of action, 106
    multiple endpoint assays, 16–17
    multiple stressors interactions, 105–106
    pathway analysis, 107
    potential contributions, genomics, 98
    real-world situations, 99–100
    regulatory monitoring, 109–110
    remediation and restoration sites, 139–140
    risk assessment framework, 89–97, *90–91*
    species extrapolation, 107–108
    systems toxicology, 112–113, *113*
    Tier III problem formulation stage, 103–105
    Tier II problem formulation stage, 101–103
    Tier I problem formulation stage, 99–100
    Tier IV problem formulation stage, 105–109
    tool development, 101–102
    uncertainty reduction, 108–109
Compound identification, 136
Comprehensive assessments, 8
Conditions and lifestages effect, 24, *see also* Lifestages
Conventional outdoor active ingredients, 68–78
Copepods, 94
Corophium spp., 94
Costs
    DNA sequencing, 21
    tiered testing, 47–49, *48*

Cross-species
    experiments, pesticides, 81–82
    human health assessments, 18
    screening applications, 26–27
    species sensitivity, 77–78
Currently used assays, 15–17

D

Dab (*Limanda* spp.), 42
Danio rerio (zebrafish)
    cross-species applications, 26, 81
    ecotoxicology application examples, 20
    model system selection, 152, 155
    organism choice, 138
    rapid response, 137
    tiered testing, 47
    transcriptomics, 22
Daphnia spp.
    ecotoxicology application examples, 20
    genome data, 3
    lifestages sensitivity, 76
    metallothionein, 101
    model system selection, 155
    surrogate species, 142
    tiered testing, 47
    transcriptomics, 41
Daston studies, 151–156
Data, 64–65
Database support application, 48, 111–112
De Coen studies, 13–27
Degitz studies, 63–83
Denslow studies, 87–113
Differential display PCR, 25
Dipstick indicators, 131–132
Direct toxicity assessment, 134–137
DNA sequencing costs, 21
Drosophila spp., 154–155, *see also* Flies
Duckweed (*Lemma* spp.), 139

E

Earthworms (*Eisenia* spp.), 43, 92
Ecological risk assessment (ERA), 19
Ecotoxicology
    fundamentals, 157
    screening applications, 22–23, *23*
Education, 110
Effects
    characterization, pesticides, 72–75
    monitoring and assessment, 92, 93–96
    tiered testing, 34–35
Effluent discharge
    fundamentals, 157
    monitoring efforts, 100
    remediation, 134–136

Index

Eisenia spp. (earthworms), 43, 92
Endpoints
　alternate assessment tools, 130–131
　exposure-response analysis, 74
　genomic cost, 49
　inclusion in workshop, 6
　multiple endpoint assays, 16–17
　pesticides, 68–69
　regulatory context, 2
　remediation and restoration sites, 140, *141*, 142
　research considerations, 53
　risk quotients, 72–73
Environmental monitoring
　fundamentals, 157
　limitations, 97
　tiered approaches, 99
　workshop, 9
Environmental risk assessment, 157
ERA, *see* Ecological risk assessment (ERA)
Euling studies, 13–27
Evans studies, 123–146
Examples
　current approaches of monitoring, 93–94
　screening, future trends, 20
Existing tools, pesticides, 81–82
Exposure assessment
　complex mixtures, 92, 93–96
　pesticides application, 70–72
　tiered testing, 35
Exposure-response analysis, 74
Extrapolation
　complex mixtures, 107–108
　confamilial species, 139
　life stages effect, 24
　species sensitivity, 78
　toxicity identification evaluation, 137

F

Fathead minnows, *see also* Fish
　contaminant mixtures, 95
　cross-species experiments, 81
　genome data, 3
　lifestages sensitivity, 76
　model system selection, 155
　tiered testing, 47
　transcriptomics, 41
Ferguson, E., studies, 123–146
Ferguson, L., studies, 63–83
Filby studies, 33–54
Fingerprinting
　cross-species sensitivity, 78
　fundamentals, 157
　future considerations, 5
　lifestages sensitivity, 75–76
　metabolomics, 43
　tiered testing, 43–44, 51–52
　transcriptomics, 43
First-generation microarrays, 25
Fish, *see also specific species*
　adverse effects linkage, 102
　cross-species, 26, 81
　ecotoxicology application examples, 20
　exposure measurements, 93
　genome data, 3
　individual sensitivity distribution, 108
　pesticides, 68–69
　proteomics, 42
　rapid response, 137
　single biomarkers, 38–40
　transcriptomics, 22, 41
Flies, 14, *see also Drosophila* spp.
Forest pests, 91
Fractionation, 95–96
Frogs (*Xenopus* spp.)
　cross-species experiments, 81
　genome data, 3
　tiered testing, 47
　transcriptomics, 41
Fundulus spp., 108
Future considerations
　complex mixtures, 110–113
　life stage effect, 24
　ongoing research, 5
　remediation and restoration sites, 138–145
　screening applications, 17–20, 25
　toxicogenomics, 17, 151–156

G

Gastropods, 92, 94
Gene expression, 39
Gene ontology, 107
Genetically modified organisms, 129
Genome, 157
Genomics
　data collection, 4–5
　development, 109–110
　fundamentals, 158
　opportunities, 97–98
　potential utility, 2
　technology incorporation, 66–68
　tiered testing, 40–44
　tools development, 82
Genotoxicity, 66
Global *vs.* focused molecular analysis, 6–7
Glossary, 157–159
Greenburg studies, 123–146
Guiney studies, 87–113

H

Haliotis spp. (red abalone), 40, 43

Heat shock protein (HSP), 39, 110
Hepatoxicity, 66
Herring gulls, 91, 92
Hoke studies, 63–83
Homeostasis, 26
HSP, *see* Heat shock protein (HSP)
Huggett studies, 13–27
Humans
    cross-species database, 26
    genome data, 3
    health assessments, 18
    metabolomics, 42
Hybridization, 25, 81, 108, *see also* Cross-species
Hyperaccumulators, metals, 129
Hypothesis-driven paradigm, 65

I

Iguchi studies, 33–54
Impacted functions, 107
Imposex
    current monitoring approaches, 94
    exposure measurements, 92
    fundamentals, 158
Individual sensitivity distribution, 108–109
Inert ingredients, 79–83
Inorganic contamination, remediation technologies, 127
Integrated Risk Information System (IRIS), 18
Integrative testing, 65–66
Intelligent testing strategies (ITS)
    genomics information, 153
    tiered testing, 36–38, *37*
    vertebrate animal reduction, 49
International efforts, 17–18
Interpretation challenges, 143
Invertebrates, 41
IRIS, *see* Integrated Risk Information System (IRIS)
ITS, *see* Intelligent testing strategies (ITS)

J

Japanese medaka, *see* Medaka (*Oryzias* spp.)

K

Kanno studies, 13–27
Kennedy studies, 13–27
Kille studies, 123–146
Klaper studies, 63–83
Kramer studies, 33–54

L

Largemouth bass, 20, 137, *see also* Fish

Large-scale demonstration projects, 83
Larsson studies, 33–54
Lemma spp., 139
Leptocheirus spp., 101
Leuciscus spp. (chub), 40, 93
Lifestages
    cross-species extrapolation, 108
    screening, priority research, 24
    sensitivity, 75–77
Limanda spp. (dab), 42
Limitations, screening applications, 20–21
Lumbricus spp., 93

M

Mammalian species, 50, *see also* Nonmammalian species
Maturity, new species, 24–25
Mechanism of action, *see also* Modes of action (MOA)
    complex mixtures, 106
    fundamentals, 158
    gene number, 3
Medaka (*Oryzias* spp.)
    cross-species experiments, 81
    genome data, 3
    metabolomics, 43
    single biomarkers, 40
    tiered testing, 47
    transcriptomics, 41
Messenger RNA, *see* mRNA (messenger RNA)
Metabolic markers, 93
Metabolites, 39–40
Metabolomics
    baseline variation, pesticides, 80–81
    cross-species sensitivity, 78
    exposure measurements, 102
    fundamentals, 4, 67, 158
    future outlook, 110
    human health assessments, 18
    lifestages sensitivity, 76–77
    limitations, 111
    mechanisms of action, 74
    monitoring efforts, 100
    potential applications, 156
    tiered testing, 42–44, 50–52
Metallothionein
    exposure measurements, 92–93
    fundamentals, 158
    future outlook, 110
    screening tool development, 101
    single biomarkers, 39
Metals, hyperaccumulators, 129
MIAME (minimum information about microarray experiment) standards, 112
Mice, *see also* Rats
    cross-species database, 26

Index

genome data, 3
metabolomics, 42
tiered testing, 47
Microarrays, 158
Minimum information about microarray experiment (MIAME) standards, 112
Miracle studies
 ecological risk assessments, 1–9
 future directions, 151–156
 remediation and restoration site applications, 123–146
MOA, see Modes of action (MOA)
Model system selection, 152, see also Organism choice
Modes of action (MOA), see also Mechanism of action
 complex mixtures, 106
 comprehensive evaluation, 153
 cross-species database, 26
 ecological risk assessment, 19
 endocrine effects, 152
 extrapolation, 137
 fundamentals, 158
 gene number, 3
 human health assessments, 18
 multiple, 16
 signaling, 40
 single endpoint assays, 16
 surrogate species, 142
 tiered testing, 52
Mollusks, 102
Monitoring, chemicals, 157
Monitoring, environmental
 fundamentals, 157
 limitations, 97
 tiered approaches, 99
 workshop, 9
Monitoring, regulatory, 109–110, 158
Mouse, see Mice
mRNA (messenger RNA)
 baseline variations, 81
 exposure measurements, 92
 multiple gene expression, 14
 single biomarkers, 39
 transcription, 3–4
Mulitphase risk assessment, 89–90, *90*
Multiple endpoint assays, 16–17
Multiple gene expression, 14
Multiple stressors interactions, 105–106
Mussels, 94

N

Near- to mid-term research needs, 65
Nephrotoxicity, 66
Nomenclature, see Glossary
Noninvasive/nondestruction sampling, 143

Nonmammalian species, see also Mammalian species
 metabolomics, 43
 model system selection, 155
 transcriptomics, 41

O

Oncorhynchus spp.
 cross-species experiments, 81
 exposure measurements, 102
 genome data, 3
 metabolomics, 43
 monitoring efforts, 100
 rapid response, 137
 tiered testing, 47
 transcriptomics, 41
Orban studies, 63–83
Organic contamination, remediation technologies, 127
Organism choice, 138–139, see also Model system selection
Oris studies, 87–113
Oryzias spp. (medaka)
 cross-species experiments, 81
 genome data, 3
 metabolomics, 43
 single biomarkers, 40
 tiered testing, 47
 transcriptomics, 41
Outcome prediction, 74
Oysters, 94

P

Pathway analysis, 19, 107
Perkins studies
 complex mixtures, 87–113
 ecological risk assessments, 1–9
 future directions, 151–156
Pesticides
 antimicrobials risk assessment, 78–79, 80
 conventional outdoor active ingredients, 68–78
 cross-species experiments, 81–82
 effects characterization, 72–75
 existing tools, 81–82
 exposure assessment application, 70–72
 fundamentals, 64–66
 hypothesis-driven paradigm, 65
 inert ingredients assessment, 79–80
 large-scale demonstration projects, 83
 metabolome baseline variation, 80–81
 near- to mid-term research needs, 65
 program development, 65–66
 proteome baseline variation, 80–81
 regulatory framework, 64–66

risk characterization, 75–78
technology incorporation, 66–68
tools development, 82
transcriptome baseline variation, 80–81
Phanerochaete spp., 129
Phenotypic anchoring
   adverse effects linkage, 102
   endpoints linkages, 6
   fundamentals, 158
   genomic cost, 48–49
   research considerations, 53
   Tier II testing, 46
   vertebrate animal reduction, 50
Phytoremediation, 128
Plaice, 137
Plant-microbe systems, 128
Populations, *see also* Model system selection
   life stages effect, 24
   remediation and restoration sites, 138–139
Potential considerations
   complex mixtures, 98
   screening applications, 19
   tiered testing, 44, *45*, 46–47
Poynton studies, 87–113
Priority research, 21–27, *22*
Profiling
   fundamentals, 158
   tiered testing, 43–44, 51–52
   triggers, 52
Program development, pesticides, 65–66
Protein expression, 15, 39
Proteomics
   baseline variation, pesticides, 80–81
   cross-species sensitivity, 78
   exposure assessment, 70
   fundamentals, 4, 67, 158
   future outlook, 110
   lifestages sensitivity, 76–77
   limitations, 111
   mechanisms of action, 74
   monitoring efforts, 100
   potential applications, 156
   tiered testing, 41–42, 44, 50

Q

Quantitative polymerase chain reaction (QPCR), 41, 158
Quantitative structure-activity relationships (QSARs)
   ecological risk assessment, 19
   inert ingredients, 79
   regulatory acceptance, 37–38

R

Rainbow trout (*Oncorhynchus* spp.)
   cross-species experiments, 81

exposure measurements, 102
genome data, 3
metabolomics, 43
monitoring efforts, 100
rapid response, 137
tiered testing, 47
transcriptomics, 41
Rapid response, 137
Rationale, tiered testing, 35–36
Rats, *see also* Mice
   contaminant mixtures, 95
   cross-species database, 26
   genome data, 3
   metabolomics, 42–43
   tiered testing, 47
REACH program, 49, 153
Real-world situations
   complex mixtures, 99–100
   genomics usage, 154
Recommendations, tiered testing, 54
Red abalone (*Haliotis* spp.), 40, 43
Registration and re-registration, 2
Regulatory framework
   complex mixtures, 109–110
   fundamentals, 2, 158
   pesticides, 64–66
   remediation and restoration sites, 125, *126*
   tiered testing, 34, 54
Remediation and resource recovery, 9
Remediation and restoration sites
   applications, 130–137
   challenges, 138–145
   chronic exposure, 139–140
   complex mixtures, 139–140
   direct toxicity assessment, 134–137
   endpoint establishment, 140, *141*, 142
   fundamentals, 124, 145–146
   future needs, 138–145
   interpretation challenges, 143
   noninvasive/nondestruction sampling, 143
   organism choice, 138–139
   population choice, 138–139
   regulatory framework, 125, *126*
   research issues, 138–145
   resources, *144*, 144–145
   risk communication, 145
   sampling, 143
   specific approaches, 131–137
   surrogate species, 142–143
   technologies, 125–129
   toxicity reduction evaluation, 134–137
Reproducibility, 20–21, 26
Research considerations
   remediation and restoration sites, 138–145
   screening applications, 21–27, *22*
   tiered testing, 50–53
Resources, remediation and restoration sites, *144*, 144–145

Index

Risk assessment
   application developments, 80–83
   complex mixtures, 89–97, *90–91*
   environmental, 157
Risk characterization, 75–78
Risk communication, 145
Risk quotients (RQs), 72–73
Roaches, 81
Robidoux studies, 87–113
Route of exposure, 105

S

Safety factors, 77, 103
SAGE, *see* Serial analysis of gene expression (SAGE)
Salmon, 47
Sampling, 143
Schaeffner studies, 123–146
Scope of workshop, 6–7
Screening applications
   biological challenges, 21
   conditions and life stages effect, 24
   cross-species, 26–27
   currently used assays, 15–17
   ecological risk assessment, 19
   ecotoxicology screens, 22–23, *23*
   examples, 20
   fundamentals, 14–15, 27
   future trends, 17–20, 25
   human health assessments, 18
   international efforts, 17–18
   limitations, 20–21
   maturity, new species, 24–25
   multiple endpoint assays, 16–17
   potential use, 19
   priority research, 21–27, *22*
   reproducibility, 20–21
   single endpoint assays, 15–16
   variability, 20–21
Screening-level risk assessments, 8
Scroggins studies, 87–113
Seasonal variations, 105
Second-generation microarrays, 25
Sensitivity, 137, 153
Sequence-poor species, 82
Serial analysis of gene expression (SAGE), 25
Sheephead minnows, 20, 137, *see also* Fish
Shellfish, 91
Single biomarkers, *see also* Biomarkers
   exposure assessment, 70–71
   tiered testing, 38–40
Single endpoint assays, 15–16
Single nucleotide polymorphisms (SNP), 24–25, 158
Snails, 26
Snape studies, 13–27
SNP, *see* Single nucleotide polymorphisms (SNP)

Spatial heterogeneity, 105
Species extrapolation, 107–108
Species sensitivity, 77–78
Specific approaches, 131–137
Sprenger studies, 123–146
Spurgeon studies, 63–83
Status monitoring, 90–91, *91*
Stepwise process, *see* Tiered testing
Steroidogenesis, 17
Stickleback, 41
Stress gene responses, 17
Stressors, 3, 105–106
Structure of workshop, *7,* 7–9
Sublethal effects
   fundamentals, 158
   monitoring efforts, 100
   multiple stressors, 105–106
Surrogate species, 142–143
Systems toxicology, 112–113, *113*

T

Technical background, 3–5
Technologies
   multiple endpoint assays, 16
   pesticides, 66–68
   remediation and restoration sites, 125–129
TIE, *see* Toxicity identification evaluation (TIE)
Tier I
   acute toxicity tests, 44, 46
   complex mixtures, 99–100
   exposure assessment, 35
   fundamentals, 35
   inert ingredients, 79
Tier II
   complex mixtures, 101–103
   current monitoring approaches, 92
   exposure assessment, 35
   fundamentals, 35
   inert ingredients, 79–80
   phenotypic anchoring, 46
Tier III
   complex mixtures, 103–105
   exposure assessment, 35
   fundamentals, 35
   inert ingredients, 80
   modes of action, 46
Tier IV
   complex mixtures, 105–109
   exposure assessment, 35
   fundamentals, 35
   inert ingredients, 80
   modes of action, 46
Tiered risk assessment, 89–90, *90*
Tiered testing
   animal reduction, *48,* 49–50
   animal species, 47
   applications, 50–53

bioinformatics, 53–54
cost benefits, 47–49, *48*
effects assessment, 34–35
exposure assessment, 35
fingerprinting, 43–44
fundamentals, 34
genomics, 40–44
intelligent testing strategies, 36–38, *37*
metabolomics, 42–43
potential applications, 44, *45*, 46–47
profiling, 43–44
proteomics, 41–42
rationale, 35–36
recommendations, 54
regulatory challenges, 54
regulatory framework, 34
research needs, 50–53
single biomarkers, 38–40
transcriptomics, 40–41
workshop, 8
Tillitt studies, 33–54
Tilton studies, 63–83
Time variations, 105
Tisbe spp., 94
Tool development
cost complex mixtures, 101–102
pesticides, 82
Toxicity identification evaluation (TIE)
contaminant mixtures, 96
extrapolation, 137
sensitivity, remediation, 137
Toxicity reduction evaluation, 134–137
Toxicity test, 159
Toxicogenomics, 159
Toxic responses, 18
Transcriptomics
baseline variation, pesticides, 80–81
cross-species sensitivity, 78
ecotoxicology screens, 22
exposure assessment, 70
fundamentals, 4, 67, 159
mechanisms of action, 74
monitoring efforts, 100
potential applications, 156
tiered testing, 40–41, 44, 50–52
Trend monitoring, 90–91, *91*
Triggers, 52
Tyler studies, 33–54

U

Uncertainty reduction, 108–109

V

van Aerle studies, 123–146
van Aggelen studies, 33–54
Van Der Kraak studies, 87–113
van Leeuwen, 33–54
Variability, screening limitations, 20–21
Versteeg studies, 123–146
Vertebrate animal reduction, 49, *see also* Animals
Viant studies, 33–54
Vitellogenin (VTG), 40, 110
Vitellogenin (VTG) mRNA, 15, 38–40
Voluntary genomic data submissions (VGDSs), 67

W

Whole-animals, *see* Animals
Workgroups, 8–9
Workshop
comprehensive assessments, 8
environmental monitoring, 9
fundamentals, 5–6, 9
global *vs.* focused molecular analysis, 6–7
overview, 5–6
remediation and resource recovery, 9
scope, 6–7
screening-level risk assessments, 8
structure, *7*, 7–9
tiered testing, 8
Worms, 14, *see also specific species*

X

Xenopus spp. (frogs)
cross-species experiments, 81
genome data, 3
tiered testing, 47
transcriptomics, 41

Y

Yeast, screening assays, 14

Z

Zacharewski studies, 13–27
Zebrafish *(Danio rerio)*
cross-species database, 26
cross-species experiments, 81
ecotoxicology application examples, 20
genome data, 3
model system selection, 152, 155
organism choice, 138
rapid response, 137
tiered testing, 47
transcriptomics, 22
Zona radiata proteins, 110